Leaf of Allah

EASTERN AFRICAN STUDIES

Revealing Prophets
Edited by DAVID M. ANDERSON & DOUGLAS H. JOHNSON

East African Expressions of Christianity
Edited by THOMAS SPEAR & ISARIA N. KIMAMBO

The Poor Are Not Us
Edited by DAVID M. ANDERSON & VIGDIS BROCH-DUE

Potent Brews
JUSTIN WILLIS

Swahili Origins
JAMES DE VERE ALLEN

Being Maasai
Edited by THOMAS SPEAR & RICHARD WALLER

Jua Kali Kenya
KENNETH KING

Control & Crisis in Colonial Kenya
BRUCE BERMAN

Unhappy Valley
Book One: State & Class
Book Two: Violence & Ethnicity
BRUCE BERMAN & JOHN LONSDALE

Mau Mau from Below
GREET KERSHAW

The Mau Mau War in Perspective
FRANK FUREDI

Squatters & the Roots of Mau Mau 1905–63
TABITHA KANOGO

Economic & Social Origins of Mau Mau 1945–53
DAVID W. THROUP

Multi-Party Politics in Kenya
DAVID W. THROUP & CHARLES HORNSBY

Empire State-Building
JOANNA LEWIS

Decolonization & Independence in Kenya 1940–93
Edited by B.A. OGOT & WILLIAM R. OCHIENG'

Eroding the Commons
DAVID ANDERSON

Penetration & Protest in Tanzania
ISARIA N. KIMAMBO

Custodians of the Land
Edited by GREGORY MADDOX, JAMES L. GIBLIN & ISARIA N. KIMAMBO

Education in the Development of Tanzania 1919–1990
LENE BUCHERT

The Second Economy in Tanzania
T.L. MALIYAMKONO & M.S.D. BAGACHWA

Ecology Control & Economic Development in East African History
HELGE KJEKSHUS

Siaya
DAVID WILLIAM COHEN & E.S. ATIENO ODHIAMBO

Uganda Now
Changing Uganda
Developing Uganda
From Chaos to Order
Religion & Politics in East Africa
Edited by HOLGER BERNT HANSEN & MICHAEL TWADDLE

Kakungulu & the Creation of Uganda 1868–1928
Michael Twaddle

Controlling Anger
SUZETTE HEALD

Kampala Women Getting By
SANDRA WALLMAN

Political Power in Pre-Colonial Buganda
RICHARD J. REID

Alice Lakwena & the Holy Spirits
HEIKE BEHREND

Slaves, Spices & Ivory in Zanzibar
ABDUL SHERIFF

Zanzibar Under Colonial Rule
Edited by ABDUL SHERIFF & ED FERGUSON

The History & Conservation of Zanzibar Stone Town
Edited by ABDUL SHERIFF

Pastimes & Politics
LAURA FAIR

Ethnicity & Conflict in the Horn of Africa
Edited by KATSUYOSHI FUKUI & JOHN MARKAKIS

Conflict, Age & Power in North East Africa
Edited by EISEI KURIMOTO & SIMON SIMONSE

Property Rights & Political Development in Ethiopia & Eritrea
SANDRA FULLERTON JOIREMAN

Revolution & Religion in Ethiopia
ØYVIND M. EIDE

Brothers at War
Tekeste Negash & KJETIL TRONVOLL

From Guerrillas to Government
DAVID POOL

Mau Mau & Nationhood
Edited by E.S. ATIENO ODHIAMBO & JOHN LONSDALE

A History of Modern Ethiopia 1855–1991
Second edition
BAHRU ZEWDE

Pioneers of Change in Ethiopia
BAHRU ZEWDE

Remapping Ethiopia
Edited by W. JAMES, D. DONHAM, E. KURIMOTO & A. TRIULZI

Southern Marches of Imperial Ethiopia
Edited by DONALD L. DONHAM & WENDY JAMES

A Modern History of the Somali
Fourth edition
I.M. LEWIS

Dhows & the Colonial Economy in Zanzibar 1860–1970*
ERIK GILBERT

*In Search of a Nation**
Edited by GREGORY H. MADDOX, & JAMES L. GIBLIN

*A History of the Excluded**
JAMES L. GIBLIN

Islands of Intensive Agriculture in East Africa
Edited by MATS WIDGREN & JOHN E.G. SUTTON

*African Womanhood in Colonial Kenya**
TABITHA KANOGO

Leaf of Allah
EZEKIEL GEBISSA

* forthcoming

Leaf of Allah

Khat & Agricultural Transformation in Harerge, Ethiopia 1875–1991

EZEKIEL GEBISSA

Assistant Professor of History
Kettering University
Flint, Michigan

James Currey
OXFORD

Addis Ababa University Press
ADDIS ABABA

Btec
HARGEISA

Ohio University Press
ATHENS

James Currey Ltd
73 Botley Road
Oxford
OX2 0BS

Ohio University Press
Scott Quadrangle
Athens, Ohio 45701

Addis Ababa University Press
& Research & Publications Office
PO Box 1176
Addis Ababa

Btec Books
Maaweel Street
Hargeisa
Somaliland
also
Btec Books
68 Kingsley Avenue
Southall
Middlesex UB1 2NB

First published 2004

1 2 3 4 5 08 07 06 05 04

British Library Cataloguing in Publication Data
Gebissa, Ezekiel
Leaf of Allah : khat & agricultural transformation in Harerge, Ethiopia 1875-1991. - (Eastern African Studies)
1. Qat - Ethiopia - Harerge 2. Agriculture - Economic aspects - Ethiopia - Harerge 3. Agriculture - Ethiopia - Harerge - History 4. Agriculture - Social aspects - Ethiopia - Harerge 5. Drug traffic - Ethiopia - Harerge 6. Harerge (Ethiopia) - History 7. Harerge (Ethiopia) - Rural conditions
I. Title
338.1'737'09632

ISBN 0-85255-481-8 (James Currey cloth)
ISBN 0-85255-480-X (James Currey paper)

Library of Congress Cataloging-in-Publication Data is available from the Library of Congress

ISBN 0-8214-1559-X (Ohio University Press cloth)
ISBN 0-8214-1560-2 (Ohio University Press paper)

Typeset in 10/11pt Baskerville by Long House, Cumbria, UK
Printed and bound in Britain by Woolnough, Irthlingborough

Contents

Photographs, Maps, Tables & Figures

Photographs

Maps

Tables

Figures

Author's Note

Ethiopian names

Ethiopians do not have family surnames. They use their first names as given names and their father's first name as last names. In the text, I will render the names of Ethiopian authors by first names. For bibliographic purposes I have alphabetized Ethiopian names according to first names.

Ethiopian calendar

The Ethiopian calendar has 365 days, with twelve months of 30 days each, and a thirteenth month of five days (six in leap years). From September 11, the Ethiopian new year, to December 31, the Ethiopian year runs seven years behind the Gregorian year; from January 1 to September 10 the difference is eight years. Wherever Ethiopian years are used in this book, the years are identified by the designation E.C.

Transliteration

There is no standard of rendering Ethiopian words into Latin script. Some use diacritics, others vowels, to reproduce pronunciation. I shall render Ethiopian words in English in a way that approximates the Ethiopian version and identify the spelling the first time the term is used.

Abbreviations

ADM	Archives d'Outre Mer
AIMK	*Association des Importeurs et des Marchands de Kath*
AMC	Agricultural Marketing Corporation
BCAF	*Bulletin de Comité Afrique*
CBD	coffee berry disease
CECCD	Chat Exporting Company of Dire Dawa
CFE	Chemin de Fer France-Ethiopien
CHATCO	Qat Importing Company
CHATHA	Khat Exporters Company of Harer
CLIA	Committees of Labor, Inspection and Audit
CO	Colonial Office
DF	Djibouti franc
E.C.	Ethiopian Calendar
ECMC	Ethiopian Coffee Marketing Corporation
EDDC	Ethiopian Domestic Distributors Corporation
EHAZ	Eastern Harerge Administrative Zone
FO	Foreign Office
f o.b.	free on board
FTAI	French Territory of Afars and Issas
FV	franco valuta
HPA	Harerge Provincial Archives
ILO	International Labor Organization
KECE	Khat Exporters Cooperative of Ethiopia
MCTD	Ministry of Coffee and Tea Development
MFT	Ministry of Foreign Trade
MOA	Ministry of Agriculture
MOI	Ministry of Interior
MT	Maria Theresa
PA	Peasant Associations
PMAC	Provincial Military Administrative Council
SD	State Department
ShS	Somali shillings
SOE	State of Emergency
SOGIK	*Société Général d'Importation de khat*
UDA	Urban Dwellers Association
WHO	World Health Organization
WPE	Workers Party of Ethiopia

Preface

This book is about khat[1] or *Catha edulis*, a psychoactive shrub, produced and marketed in the eastern portion of an expansive plateau located in the province of Harerge, Ethiopia. Focusing on the period between 1875 and the early 1990s, this work attempts to reconstruct the history of the people who rely on khat production for their existence; and tries to answer the questions of why farming families over the last half-century increasingly shifted from cultivating coffee and food crops to growing khat. Here the study seeks to show how demographic, market and political factors facilitated the emergence of khat as Harerge's leading agricultural commodity and how this development, in turn, increased the scale of unofficial cross-border trade in consumer goods. The study also explores the consequences of the new cash crop for the regional economy as a whole, for farmer–state relations, for the nature and balance of local social relations, as well as for Harerge's physical, socioeconomic and political landscapes.[2]

Before 1900 agriculture was the main economic occupation of the residents of Harer, the largest and oldest urban center in the region. Though city-dwellers, the townspeople maintained gardens and cultivated fruits, coffee, khat, spices, vegetables and cereals such as sorghum, maize, wheat and barley. Coffee was the most important of the agricultural commodities exported from Harerge. Khat was cultivated mainly by the Hareri for local markets. Only dried khat was exported to the neighboring countries until the Addis Ababa–Djibouti railway track reached the foothills of Harer in 1902.

At the turn of the twentieth century, the consumption of fresh khat for pleasure was limited to a small, well-off urban elite. Written and oral sources suggest that the main users were religious leaders who chewed the leaves to stay awake and pray; long-distance traders who preferred to travel at night; and farmers who chewed it for energy. Since the 1940s, however, the demographic configuration of users has changed dramatically. In contemporary Harerge, only a tiny minority of the population, for personal

[1] I have counted at least fifty names for khat. Although farmers in Harerge use the Oromo word *jimaa*, I will use the term khat because it is widely used and known. For details, see P. Greenway, 'Khat,' *The East African Agricultural Journal*, 13 (1947); Bertil Nelhans, *Khat: A Stimulating Drug in East Africa and the Arabian Peninsula* (Gotenborg, 1974), pp. 3–5.

[2] There have been a number of changes in the administrative divisions of Ethiopia in recent years. In this book, unless otherwise indicated, Harerge denotes the province that existed until 1987, and Harer refers to the provincial capital.

and medical reasons, refrains from chewing khat. Daily laborers, construction workers, farmers, Quranic readers and reciters, students, long-distance truck drivers and cabbies, animal hunters, couriers, Muslim judges, men, women and children, all chew khat.

The dramatic expansion in khat consumption has been accompanied by a significant increase in its economic importance. Responding to the growing demand and the consequent price-rise for the product, a considerable number of farmers in the region have abandoned the production of other cash and food crops and transformed their plots and gardens into khat fields. The new, easy-to-handle cash crop allowed the poor agrarian population to fulfill its civic obligations, acquire its basic necessities and make productivity investments. Residents of urban centers in the region also profited from participating in the khat economy as retailers and as operators of the cross-border trade. Many changes occurred during the period under consideration, and much khat-generated capital was invested in nonfarm sectors, but khat continued to be the focal point of the region's agriculture and, perhaps, the main source of wealth for the regional mercantile class. By the end of the 1980s, khat production and marketing had made the Harerge province an exemplar of prosperity and economic development in an otherwise poor country. Without a doubt, khat is currently the fastest-growing and the most profitable occupation – involving millions of farmers, traders and other auxiliary service providers – in the whole region.

The research for this book was conducted in Harerge in 1994 initially for a Ph.D. dissertation, which I presented to the history department of Michigan State University in 1997. A comment by one of my committee members criticized my data as reflecting only the view from the khat production side, Ethiopia. Indeed, all of my oral materials, unpublished sources and archival records, whether written by Ethiopians or expatriates, were generated there. Most of the archival materials I consulted at the Public Record Office in London told the story of khat from the Ethiopian side. Even oral data seemed to lack coverage, since it was collected from about a hundred informants. I knew then that I needed to return to Harerge to conduct intensive interviews, as opposed to the extensive ones I had done in 1994; and to Europe to look into archival materials for a view from the khat-receiving side, the Aden Colony and the Somaliland Protectorate.

In January 2000 I went to Ethiopia for a month of follow-up and to catch up on the changes in Harerge stimulated by khat's continuing rise in importance. I was surprised by the broad interest in khat, both in controlling it, owing to its economic importance, and in studying its impact on society and on individuals. In Addis Ababa University, one M.A. thesis in anthropology had been completed; newspapers had published a series of stories on various aspects of khat; radio and television journalists had conducted and were conducting interviews with specialists and government officials on the subject. On one occasion, I read the comments of the Deputy Prime Minister of Ethiopia that, in 1999, khat had become the second largest foreign exchange earner for the country. The volume of the exported khat increased by a staggering 84 percent and the value by 65 percent over the previous year.[3] No scientific surveys exist of the proliferation of khat retail shops in the capital, Addis Ababa, or in any town in the country, but even the casual observer cannot fail to witness that there are more khat stalls and shops in Addis Ababa than in the past and that there is an increased enthusiasm for khat consumption on the part of all generations and classes of Ethiopian society.

But research for a book cannot be undertaken solely on the basis of personal observations and anecdotal evidence. In Harerge, I selected two individuals whose lives were wholly transformed owing to their association with the khat trade. I interviewed these individuals for two days each, in most cases listening to their stories with as little interference from me as possible. Theirs are fascinating life histories that speak a great deal about khat's present place and role in Harerge's society and economy. Their vignettes are interspersed throughout the narrative to provide a human dimension to the quantitative data I have used to document the dramatic changes khat has initiated in Harerge. I also collected local newspapers that have reported on various aspect of khat use and market and local studies carried out by experts in the field who worked for the government agencies responsible for agricultural and commercial affairs. Most of the newspaper reports and local studies pertain to a period that falls beyond my study, but they are very useful indicators of the continuing trend with regard to the expansion of production and marketing.

I also returned to the Public Record Office in London to read through reports on khat from Aden Colony and the Somaliland Protectorate covering the period from 1946 to 1959. Officials from the British territories took great interest in khat, perhaps bewildered by the local culture or concerned about the effects of the strange plant on their subjects. Various official documents such as annual reports, commissions of inquiry reports as well as official correspondence on the subject of khat flowed from Hargeisa and Aden to the Colonial Office. These documents largely reflect British attitudes toward khat, but they also provide important insights into a very critical period in the commercial relations between Harerge, the Aden Colony and the Somaliland Protectorate.

The follow-up research evened out the lopsidedness of my sources by incorporating the view from the receiving side, Aden and Hargeisa. The oral sources, archival materials, government and empirical studies were used to construct an adequate answer to the main question of this book: why did farmers in Harerge switch from coffee and food-crop cultivation to khat production? The available evidence has allowed me to identify the factors that initiated the expansion of consumption, production and marketing, and the overall change in the political economy of Harerge wrought by khat's commodification. It was also possible to explore broader issues of agrarian change, the economics of unofficial cross-border trade and society–state relations. I hope and expect that this book will provide some clarity and insight to those who make policy decisions that affect

3 'Khat has become the backbone of our country's economy,' *Moresh* (Amharic weekly), 93, January 24, 2000.

millions of farmers and traders in Harerge and in the rest of Ethiopia.

Many individuals and institutions assisted me at various stages in the preparation of this book. I am grateful to the Rockefeller Foundation for enabling me to undertake fieldwork and archival research in Ethiopia, Great Britain and France. Kettering University set aside sufficient funds for my research through its New Faculty Professional Development Grant program. The money enabled me to travel to my research site to do more interviews and to Europe to conduct archival research. I am indebted to Kettering University for this and for its continued support for research.

This book would not have been possible without the support, cooperation and friendship I received from so many individuals over the years. I owe a great deal of intellectual debt to the late Harold Marcus, my mentor, who became my friend and colleague. He met me when I was a young historian working in the National Archives of Ethiopia. He invited me to come to Michigan State University for graduate studies and supported my work throughout my graduate career. I thank Prof. Marcus for encouraging me to be imaginative, critical and creative, and hope that I have lived up to his expectations. I also want warmly to acknowledge the intellectual guidance and support of Professor David Robinson, who not only taught me the art of doing history, but also demonstrated to me the value of discipline, persistence and diligence. Professors Elizabeth Eldredge and Martin Daly gave me valuable suggestions and comments, which I gratefully acknowledge.

I am also deeply indebted to the employees of the provincial archives in Harerge for carrying dusty files to me day after day, and to the workers of various government agencies in Harer and Dire Dawa for their generous assistance and warm hospitality. I would like to recognize the support of Samya Yusuf, Head of the Dire Dawa Foreign Trade Office, Abebe Mekonnen and Ashenafi Belayneh of the Dire Dawa and Eastern Harerge Offices of Economic Development and Planning respectively, and Dr Milkessa Wakjira of Alemaya University of Agriculture. I also received courteous and informed assistance from the staff of the Institute of Ethiopian Studies, Addis Ababa University. Above all, I will always be indebted to my informants in Harerge, who accepted me in their midst and endured with grace my endless questions. I hope this book reflects their hopes, aspirations and quest for a better life. I also wish specifically to thank the Rev. Mezgebu Fufa and his family for their loyal friendship, boundless generosity and moral support. My brother, the Rev. Barnabas Daniel and his wife, Hirut, always welcomed me in their home in London and made my stay there enjoyable, productive and memorable.

Finally, I would like to thank my wife, Meseret, who had to endure with me the challenges of balancing the high-pressured life of a professional couple with the daunting task of raising two energetic children, Haara and Kenna, who have made it possible for me to find joy in both work and family. I owe all of them much more than I can ever say.

East Lansing, Michigan

Part One

Beginnings

One

Introduction
Social & Historiographical Contexts

> Khat is a tree that God loves. It's a tree blessed by Rabi [God] and given to us. This is a tree that man cannot command. A lot of people with power have tried to control it, especially the price of selling the leaves. None has succeeded so far. This tree is not just another ordinary plant; it's a 'Leaf of Allah.'
>
> Adem Aliyu Jaylan, 2000[1]

Khat, whose leaves and twigs are chewed for their euphorizing effects, is a popular stimulant known and used in many parts of the Middle East and the Horn of Africa for at least a millennium. In the east-central part of Ethiopia, chewing khat goes back almost eight hundred years to the onset of the Christian–Muslim struggle in medieval Ethiopia. In a contemporary account of Emperor Amde Syon's (r. 1314–44) military campaign undertaken against Sebra'din of Ifat around 1330, the latter is quoted as saying, 'I will take up my residence at Mar'ade, the capital of his [Amde Syon] kingdom, and I will plant chat [khat] there because the Muslims love the plant.'[2] The story suggests that the use of khat was a familiar practice among Muslim communities in east-central Ethiopia around the mid-fourteenth century. Ever since, khat has been a fundamental part of the social and cultural experience of the people of Harerge. The importance of this plant cannot be overstated because it still plays a vital role in the lives of millions of Oromo farmers who constitute the majority of Harerge's rural population and in the lives of various ethnic groups residing in the province's urban centers.

[1] Interview: Adem Aliyu Jaylan, 61, AwWoday, January 19, 2000.

[2] The chronicle has been translated into English by G.W.B. Huntingford, *The Glorious Victories of 'Amda Syon, King of Ethiopia* (Oxford, 1965), pp. 55–6; khat was also mentioned in a fourteenth-century work by Ibn Fadl Allah al-'Umari (1301–48), who refers to its popular consumption in Ethiopia and describes its physiological effects: Ibn Fadl Allah al-'Umari, *Masalik al-absar fi mamalik al-amsar*, traduit et annoté avec une introduction par Maurice Gaudefoy-Demombynes (Paris, 1927), pp. 11–13.

While Ethiopia is the leading producer and exporter of khat, the plant is grown in Kenya and to a lesser extent in Tanzania, Malawi, Uganda, the Democratic Republic of Congo (Zaïre) and South Africa. The Meru district of Kenya has long owed its existence to a thriving khat commerce.[3] Tanzania and Uganda attempted with some success to reduce their khat crop and curtail domestic consumption in the 1950s.[4] Although khat has been more prominent in the Harerge highlands and has become the focus of most present-day attention, the plant is grown throughout Ethiopia, from the Sudanese border in the west to that with Kenya in the south.[5] In Asia, although khat has been known to grow in and around Afghanistan and has been successfully cultivated in India and Sri Lanka, the highlands of Yemen have been the main region of khat cultivation for a long time.

Until the enormous growth of fresh khat consumption after the Second World War, as we shall see later, the primary consumers of commercial khat in Harerge were the urban residents of the cities of Harer, the cultural and political capital of the region, and Dire Dawa, the commercial center since the turn of the century. As the center of the khat culture, the Harerge region gained notoriety earlier in the twentieth century. Many critics were among the Christian Amhara rulers and settlers in Harerge who established suzerainty over the local Muslim Oromo, Somali and Hareri peoples in the late nineteenth century. Representing themselves as benefactors of the conquered natives, they condemned khat use as a scourge of society and urged its proscription.[6] Outsiders' perception of khat has been negative for a long time. Travelers' accounts, official publications or scientific research invariably cast khat as a harmful, addictive drug. The attitudes of national policy-makers and officials of international organizations toward the plant did not escape being colored by the negative images.

In contemporary Harerge, the Muslim Oromo chewers describe the experience of chewing khat in glowing terms while some sectors of the province's population continue to be critical of its use, carrying on the long-standing debate on the value or potential harm of the indigenous uses of khat. Over the last half century, the social and religious strictures against khat-chewing have collapsed and khat cultivation and use have

[3] Known as *miraa* in Kenya, khat has been one of the most important cash crops in the Meru and Embu districts and probably the most important one in the Nyambeni Hills. See Anders Hjort, 'Trading Miraa: From School-leaver to Shop-owner in Kenya,' *Ethnos*, 39 (1974), pp. 27–43.

[4] CO 822/1824, Use of Khat (Qat) Kat in East Africa.

[5] Khat has probably grown in many parts of Ethiopia for centuries. A French scientific expedition led by Lieutenant Charlemagne Théophile Lefebvre, undertaken in 1839–43, confirms that the shrub grew profusely in Ethiopia's Tigre and Shewa provinces. Théophile Lefebvre *et al.*, *Voyage en Abyssinie exécuté pendant les années 1839, 1841, 1842, 1843* (Paris, 1845–51), pp. 134–5.

[6] The attempts in Ethiopia to eradicate khat recall the European effort to outlaw khat use in their colonies. The British banned khat in the Somaliland Protectorate in 1921 and their Aden Colony in 1958. The French unsuccessfully tried to do the same in Djibouti in 1957.

expanded substantially. The majority of the residents of the Harerge region, regardless of their religious and ethnic affiliation, now acknowledge more favorably the place of khat-chewing in contemporary society as a powerful symbol of cultural identity. Indeed, the most recent scholarship on khat's role in society is marked by sensitivity to the cultural context of widespread consumption of the leaves. Its main feature has been the argument that people's enthusiasm for khat consumption must be explained in social and cultural terms rather than in purely biological ones, which have consistently presented chewers as helpless addicts.

The latest scientific works have gone a long way toward correcting the many erroneous assertions about khat as an addiction-producing drug with deleterious consequences for the chewer.[7] Ironically, a parallel conception of khat use in the opposite direction has developed in some parts of the world and in some regions of Ethiopia. France and recently the United States have decided to ban the importation of khat on grounds that it is an addictive drug. It is legal in Britain, Israel, Canada, the Netherlands and Yemen.[8] Said Barre outlawed it in Somalia in 1983, but the prohibition has proved unsuccessful.[9] In light of this confusion it is necessary that we have a better understanding of the cultural aspects of traditional khat consumption in Harerge.

Khat consumption in Harerge

Owing to its stimulating effect, khat has long been popular with Islamic scholars, farm laborers and long-distance travelers in the Red Sea region. For many of these users, chewing the leaves has been the most prevalent method of consumption. Fresh khat leaves, two or three days off the tree at the most, which have kept their moisture, were, and still are, appreciated by chewers because the more tender and juicy the leaves, the easier they are to chew and more likely quickly to produce the desired state of euphoria. When fresh khat leaves were unobtainable, either a kind of paste made from powdered leaves would be ingested or an infusion made by boiling the dried leaves would be drunk. In the

[7] The best-known recent study to promote a more positive image is Shelagh Weir, *Qat in Yemen: Consumption and Social Change* (London, 1985). Kevin Rushby, *Eating the Flowers of Paradise: A Journey through the Drug Fields of Ethiopia and Yemen* (New York, 1999) is less academic but presents a fascinating picture of the culture surrounding khat in Harerge and Yemen. The most comprehensive study on the science of khat is John G. Kennedy, *The Flower of Paradise: The Institutionalized Use of the Drug Qat in North Yemen* (Dordrecht, 1987).

[8] Khat is placed in Schedule I of Controlled Substances in the US, the most dangerous category. It is illegal, to my knowledge, in Saudi Arabia and Tanzania. In Ethiopia, two major administrative zones have banned khat. Editorial, 'Despite War, Famine and Pestilence, the Khat Trade Thrives in East Africa,' *The New York Times*, December 14, 1995, p. A8; Kevin Rushby, 'The High Life,' *Geographical Magazine* (January 1995), pp. 14–17.

[9] Abdi Ismail Samatar, *The State and Rural Transformation in Northern Somalia, 1884–1986* (Madison, 1989), pp. 145–7.

absence of rapid means of delivery, khat was necessarily consumed dry. Very rarely was khat smoked as was tobacco or cannabis, although the ends of the twigs and leaves were sometimes crushed and rolled in cigarettes.[10]

From the 1840s on, we find constant references to khat and its use in various parts of Ethiopia. In 1843 Charles Beke reported that coffee and khat were growing wild in southern Abyssinia. Citing Carsten Niebuhr, Beke also identified khat as a favorite Arab intoxicant used as some sort of tea, and suggested that an infusion was made out of khat and consumed as a beverage.[11] From this information, one could infer that the practice of chewing khat was not widespread in those parts of southern Ethiopia where Beke visited in the mid-nineteenth century.[12] However, the use of khat was widely practiced in central Ethiopia during William Cornwallis Harris's visit to the court of King Sahle Sellassie (r. 1813–47) of Shewa in 1841. In his *The Highlands of Aethiopia*, Harris wrote that, in Ifat, the fresh leaves of khat were chewed and used as an astringent medicine, boiled with milk or water, and drunk as a beverage. He also spoke of the leaves being plucked and dried in the sun.[13] It is probable that the dried leaves were meant for use in regions distant from the production areas.

One of the more widely cited works on mid-nineteenth-century Harer is Richard Burton's *First Footsteps in East Africa* (1853). A keen observer and a fine reporter, Burton characteristically provided a detailed and insightful account of a khat-chewing party in Harer. He noted that such events took place every day between 9 a.m. and noon, and he described one at the home of the Treasurer of the Emir of Harer in which several dignitaries, including the prime minister, participated. At this chewing session, fresh khat leaves were pounded in a wooden mortar and the paste was distributed among the participants, who rolled the substance into chewable bales. The method of choice, however, was to pluck off the tenderest leaves and then chew. Also recorded in Burton's account was the ulema's notion that khat was 'food for the pious' and an intellectual's rationalization that it had special properties of 'enlivening the imagination, clearing the ideas, cheering the heart, diminishing sleep, and taking the place of food.'[14]

By the time the Egyptians occupied Harer in 1887, chewing was a commonplace practice among pious Muslims of the city and its environs. Mohammed Moktar, an officer in the Egyptian army, gives a most detailed account that is worth quoting at length:

[10] UN Commission on Narcotic Drugs, 'Khat,' *Bulletin on Narcotics*, 8 (1956), p. 11.

[11] Carsten Niebuhr, *Travels through Arabia and other Countries of the East*, translated into English by Robert Heron, vol. 2 (Dublin, 1792).

[12] Charles Beke, 'On the Countries South of Abyssinia,' *Journal of the Royal Geographical Society*. 13 (1843), p. 263.

[13] W. Cornwallis Harris, *The Highlands of Aethiopia* (London, 1844), vol. 3, pp. 334–5.

[14] Not known in other khat-chewing areas, distribution of khat paste is perhaps peculiar to Harer. Richard F. Burton, *First Footsteps in East Africa or, An Exploration of Harrar*, vol. II (New York, 1987), p. 31.

> As an intoxicating substance they [the Hareri] chew the leaves of a tree which they call kat. They believe that this tree has the ability to strengthen the body, to repel sleep and they attribute aphrodisiac qualities to it. They have a curious manner of using it and it seems to me that it would be of interest to relate it here.
>
> Towards nine o'clock in the morning, all the guests go to their hosts; there they sit in a circle and begin to read the first chapters of the Quran, and address all sorts of praises to the Prophet. This done, the master of the house gives each a fistful of kat leaves which they chew in eager rivalry, in order to be able to swallow them more easily. If the master of the house is rich, they drink milk; if he is poor, they drink water. After this the same ceremony begins again, reading the Quran, praises to the Prophet, receiving and chewing of a fresh fistful of kat, and this goes on until 11 o'clock. As I asked one of them why they were reading the Quran in this way and celebrating, with the eating of kat, the praises to the Prophet, he answered me.
>
> 'We read the Quran and we bow to the Prophet because this plant is known to the saints and it permits us to keep vigil long through the night in order to worship the Lord.'
>
> I had to admit that the response was rather ingenious... . I do not know the real properties of kat; this is certain, that Diabe, my servant, aged 23 years, and of a sanguine temperament, fell sick with dysentery, and was only cured with kat,... I learned myself that it was an excellent remedy against dysentery.[15]

Moktar's description shows that, by the 1880s, khat-chewing and the accompanying rituals had assumed the characteristics currently observable in Harer. It also reveals that khat was used as medicine and that chewing was closely associated with religious observances. From the nineteenth-century sources emerge three important factors still relevant to understanding contemporary khat culture. First, khat-chewing is indeed a longstanding practice in the Muslim-inhabited regions of east-central Ethiopia. Second, khat consumption by the masses of Harerge is a fairly recent phenomenon. As late as the closing decades of the nineteenth century, most of Harerge's urban dwellers, lacking the wherewithal of the affluent, chewed khat only on special social and ceremonial occasions, as did the residents of rural areas. Third, as the following description of contemporary patterns of use shows, the method and style of khat-chewing have not changed in the last one hundred and fifty years.

Khat's social & cultural aspects

Any discussion of the nature of contemporary khat consumption in Harerge must begin by distinguishing between perceptions in Harerge's rural and urban areas. For the Muslim Oromo farmers of rural Harerge,

15 Mohammed Moktar, 'Notes sur le pays de Harrer,' *Bulletin de la Société Khédiviale de Géographie,* 1:5 (1876), pp. 369–72.

khat is valued for its critical role in such productive activities as work, meditative worship and cultural ceremonies. For users in cities and towns, khat is a recreational substance used (or more precisely abused) by young men and women solely for social pleasure. Interestingly, what is common to both categories of chewers is the profound cultural importance of khat as a medium of social interaction.

Khat leaf is heavily consumed in the rural areas of Harerge by Muslim Oromo farmers. A typical farmer commences the day by going into his *oyiru*, the family garden, to feed his animals and inspect the farm. Between eight and nine o'clock, he walks into his khat orchard and settles down with other men of the family or the neighborhood for the morning chew, the *igabana*, literally the eye-opener. This session usually lasts about an hour. Only a small quantity of khat is chewed at this time for the purpose of it all is to achieve a quick *mirqaana*,[16] the desired state of euphoria, for a burst of energy for the work ahead.

The chew session is followed by a breakfast, brought to the field by the farmer's wife. For the next three to four hours, the farmer works his farm vigorously, preparing seed-beds for planting, or shoring up the khat tree, watering and fertilizing or pruning and defoliating, depending on the season. Rural chewers suggest that, in addition to the focused mind-set and increased concentration, *mirqaana* gives them the physical energy they need for their labor-intensive daily activities. They reject the critics' charge that chewers have a lazy attitude toward work.

In the early afternoon, another chewing event starts, the *bartcha*, a popular and commonplace sight throughout the region. Chewing takes place inside or outside the home, depending on the location of the chewer's dwelling. In rural areas, small groups of farmers assemble at various places near or in khat fields and settle down under a shade to rest, converse and chew the leaf for an hour or two. Urban users congregate in a designated room in the house and lie on their sides against a pile of cushions to meditate, read and engage in the talk of the town. Most people bring their own khat to the routine daily *bartcha*. Farmers may obtain their supply from their *oyiru*, but urbanites generally purchase their *zurba*, one bundle, of khat in the *chat terra* or the town's daily market. The host provides the leaves if the *bartcha* was an invited gathering to observe a given ceremony, accomplish a task or merely get together.

Once in place, the chewers take the leaves from a plastic bag or a

[16] Popular stories abound in Harerge concerning the behavior of chewers in a state of *mirqaana*. The stories were so pervasive and fascinating that a programmer of the state-owned Radio Ethiopia decided to produce and air a radio documentary on the subject. One of the stories has it that a man was chewing khat alone in his home. As most chewers do, the man plucked off the tender shoots and leaves to chew and threw away the leathery leaves. The unwanted parts piled up and covered most of his feet except for the toes, which he wiggled unconsciously. The chewer was charged with the effects of the khat and his vision was blurred to the extent that he mistook his toes for a mouse. Scared, the man quickly grabbed a machete and chopped off his toes, only to realize after the high wore off that he had mutilated himself. Interview: Ahmad Tawfiiq Ibrahim, Dire Dawa, January 14, 2000.

banana-leaf wrap, remove the tender shoots and place several leaves, one at time, in their mouth. The leaves are positioned between the inner side of the cheek and the gum on one side of the mouth. Khat-chewing is always accompanied by the drinking of water or an infusion of coffee husks or milk in order to reduce dryness of the mouth, soften the leaves and assist the extraction of the juice. The chewers continue to add leaves until a quid forms, which is sweetened by a pinch of sugar or a sip of Coca-Cola to moderate the bitter taste. Preparing the full quid may take as long as thirty to forty-five minutes. Most urban and some rural chewers smoke cigarettes as they chew. At the end of the *bartcha* the wad is usually expectorated, but most experienced chewers swallow it with water, tea, milk or pop, as a final act of the ceremony. This stage of the elaborate chewing ritual is called *luluqacha*, or rinsing the mouth.

At this point, the ritual of *bartcha* in rural and urban areas diverges. Rural chewers usually put their *mirqaana* or their heightened energy to work. Farmers and farm workers head back to the fields to work until dusk, at which time the chewers eat dinner, usually a plentiful quantity of chicken with rice or pasta. In light of the critics' perpetual contention that khat-chewing fosters malnutrition by ruining the appetite, it is worth pointing out that, in Harerge's rural areas, khat is chewed before or after meals. No one who has actually observed the customs regulating khat-chewing could sustain the 'food deprivation' hypothesis.

In an urban context the *bartcha* session is usually followed by a drinking binge of alcoholic beverages called *chabsii*, literally to break [the *mirqaana*]. Urbanites argue that alcohol counteracts the inebriating effects of khat and allows the chewer to sleep well and rise the next morning for work. Since most urban chewers are office workers or private businessmen, their explanation does make sense, even though the practice is considered anathema to the rural chewer. The Muslim Oromo farmers find the notion that alcohol is an essential ingredient of the khat-chewing ritual virtually incomprehensible. In their view, mixing the sacred khat with the defiling alcohol violates all the norms and etiquettes associated with traditional khat-chewing. 'This is a waste of time and money,' retorted one of my informants. 'Why would anyone want to swing between *mirqaana* and *harrara* [a depressed state of mind and physical weakness resulting from lack of khat]? It makes no sense.'[17] In the 1960s one researcher was told by Oromo farmers: '… if cat [khat] chewing is not followed by hard labor it then serves as an irritant rather than as a stimulant.'[18] It is evident that rural users have long disapproved the

[17] Interview: Ahmad Tawfiiq Ibrahim, Dire Dawa, January 14, 2000. When the *harrara* sets in, most habitual chewers are said to behave erratically. My informant relates that one such habitual chewer went to bed one night without chewing khat. While asleep, a ghost appeared to him, and picked him up by his hair and threatened to drop him in a bottomless pit for failing to chew khat. The recalcitrant chewer promised the ghost that he would chew khat if it spared his life. The ghost then gave the chewer a bunch of khat, which the latter began to chew immediately. When the chewer woke up in the morning, he realized that he had taken a big bite out of his blanket.

[18] Bob Hill, 'Cat (*Catha edulis* Forsk.),' *Journal of Ethiopian Studies*, 3 (1965), p. 14.

urban style of khat use. The *bartcha* has nevertheless remained a recreational practice for the urban user.

While it is not unusual for some rural chewers to take some khat before bed time, evening or night-time chewing is not regular and widespread. This last chew session of the day is known more appropriately as the *atarora,* even though some city folk mistakenly refer to it as an all-night *bartcha.* The Muslim Oromo of rural Hererge do engage in an all-night chew session, in most cases, during the month of Ramadan, but they differentiate the *atarora* from *bartcha.* The Ramadan all-night session is considered a solemn occasion of reciting the Holy Qur'an, performing prayers and meditative worship and is different from the brief chew before going to bed.

Regardless of where and when they take place, the *igabana*, *bartcha* and *atarora* provide the social context within which peaceful and constructive interaction takes place. An invitation to chew khat is a sign of the participants' commitment to rational and peaceful conversation. While solitary khat-chewing is not customary, for someone working alone, a brief *bartcha* provides a meditative interlude in which to prepare mentally for the task ahead.

The meaning of khat-chewing

In Harerge khat has historically served the same function as such mild stimulants as kola in West Africa[19] and coca among the indigenous people of South America.[20] It is chewed to reduce fatigue and sensations of hunger and thirst – the leaves are extremely bitter, probably the property that relieves thirst. Contemporary users explain that political and religious leaders customarily gave visitors khat as a sign of politeness, while individual hosts have been expected to provide a handful of the leaves to friends and guests as an indication of hospitality. Among the Oromo, births, marriages and religious festivals are celebrated by chewing khat and, on occasions such as funerals[21] and naming ceremonies, khat leaves are presented as gifts. During the popular *wadaja* ritual – a ceremony of group prayer performed at times of illness, death or calamity – large amounts of khat are consumed by the participants.

Khat has long been an integral part of all social gatherings of significance, ranging from a friendly get-together to community obligations such as weddings and funerals. Even today, friends and relatives bring khat to a *taziyya*, a mourning, as a way of consoling the bereaved. Khat leaves are furnished at weddings to add glow and glitter to the occasion and increase felicity for the newly-wed couple and the invited.

[19] Paul Lovejoy, *Caravans of Kola: The Hausa Kola Trade, 1700–1900* (Zaria, 1980).

[20] Catherine J. Allen, 'To be Quechua: The Symbolism of Coca Chewing in Highland Peru,' *American Ethnologist*, 8 (1981), pp. 157–71.

[21] Andargachew Tesfaye, 'The Funeral Customs of the Kottu of Harer,' *University College of Addis Ababa, Ethnological Society Bulletin*, 7 (1957), pp. 37–9.

When there is no occasion to celebrate, khat commits an individual to participate constructively in fairly casual social situations by providing the most comfortable atmosphere for several hours of intimate and useful conversation. In all of these cases, khat serves as an essential social lubricant that fosters amity, cooperation and sociality.

As a social institution, khat is used to seal important and long-term contracts. For example, the use of the leaves in dowry arrangements is evident in contemporary Harerge. When an Oromo couple enters into a *naqata*, an engagement, the groom's parents send *jarsaa* or representatives to negotiate the matter with the bride's family. The first act in this process is the presenting of *jimma naqata* or 'engagement-khat' to the young woman's parents, who will signify their agreement through accepting the leaves. Refusal would be interpreted as a polite rejection of the proposition.

The presence of khat in worship practices further demonstrates how the leaf has become institutionalized in the religious life and practices of the Muslim population of Harerge. Khat is highly regarded by Muslims, who report that the plant was a *kutamomia* (that which is blessed by God and given to men through His favor).[22] For centuries, it has been a standard practice for those who participated in religious ceremonies held at the Muslim shrines in the province to spend long hours of the day and night chewing khat while reciting passages from the Holy Qur'an and praying to Allah. Chewing enjoyed an important role during Ramadan and at the *Arafa* celebrations, which are particularly popular in Harer. Religious readings and meditations are preceded by chewing khat during the *Mawlid*, the birth day of the Prophet.

Emphasizing the critical cultural role and meaning of khat in Harerge, one historian summed up the rationale for khat-chewing as: 'the Oromo, the Adere, the Afar and Somali counted it [khat] among the essentials necessary for living, work, and enjoyment. Guests are welcomed with it; prayers were kept long and lively with it; in wedding and funeral ceremonies as well as other social gatherings people were supplied with bunches of the leaves as a matter of course.'[23]

Factors determining the scale of khat consumption

In the nineteenth century, affordability was probably one of the reasons why khat consumption remained the luxury of the financially well-off

[22] Clarke Brooke, 'Khat (*Catha edulis*): Its Production, and Trade in the Middle East,' *Geographical Journal*, 126 (1960), pp. 52–3. Demissie Gebre Michael, *The Kottu of Ethiopia: An Introduction to the Eastern Oromo*, Imperial Ethiopian College of Agriculture Experiment Station Bulletin No. 68 (Dire Dawa, 1974). Brooke and Demissie were told by Oromo informants more or less the same story regarding the social and religious significance of khat in Harer during the time of their research in the early 1950s and early 1970s, respectively. Robecchi-Bricchetti also reported a similar story seventy years before Brooke: L. Robecchi-Bricchetti, *Nell'Harrar* (Milano, 1903), p. 133.

[23] Bairu Tafla, 'Addictives in North East Africa: A Brief Survey of an Aspect of Nineteenth Century Cultural History,' *Afrika und Ubersee*, 64:2 (1982), pp. 282–309.

Photo 1.1 *(top)*
Khat-chewing while worshipping Allah

Photo 1.2 *(left)*
Guests are welcomed with khat (the author is standing second from left)

within the outskirts of the city of Harer. Later, the area of khat consumption was determined by the perishability of the leaves and location of supply and demand. Fresh leaves are preferred by consumers, since the desired mental state is more quickly achieved when khat is consumed within forty-eight hours of being cut. This longstanding belief of consumers is also supported by scientific research that the main active ingredient in khat, cathinone, degenerates to the less potent cathine, which is found in dried leaves. In the absence of rapid means of transportation, fresh khat could not have been exported to far-off places. The consumer pool consequently remained small until the arrival of the Djibouti–Addis Ababa railway at Dire Dawa in 1902, the construction of highways during the Italian occupation of Ethiopia (1936–41) and, finally, the introduction of air-transport after the Second World War. These events opened vast and new markets for khat and began a period of a steady expansion of consumer demand.

While improvements in the system of transportation obviously made fresh khat readily available in regions where it had rarely been present, many people in Harerge took up the practice for different reasons. For instance, the Oromo accepted the culture of chewing with their conversion to Islam around the turn of the century. In the 1940s and 1950s, many other people found employment in the expanding civil service and economy, which provided them a decent income and surplus cash to spend on luxury products such as khat. Consumption continued to grow in the 1960s, gradually but steadily, as khat became the stimulant of choice for high school and college students, who chewed the leaves during hours of studying for exams and for recreation in a region with limited entertainment amenities.

A massive change in attitude toward chewing occurred and the habit spread more rapidly and widely during the period Ethiopia was ruled by the *derg* (literally committee), the military government of Ethiopia from 1974 to 1991. The reasons for the phenomenal expansion during the *derg* years are complex and not always easily explicated, but the increased politicization of the youth in the 1970s and the participation of all high school and college students in the so-called *zemecha*, the Campaign for Development through Cooperation, played a significant part in the sharp rise of the number of khat consumers. During the *zemecha* years (1975–7), the majority of the students were sent to rural areas to implement the Land Nationalization Act of 1975 and to heighten the political consciousness of farmers. There, away from the immediate supervision of parents and the enforcement of community norms, the students took to khat consumption, along with cigarettes and alcohol. When they returned from the *zemecha*, the students brought back to their urban settings the culture of chewing and conversation.

In the absence of a systematic survey it is difficult quantitatively to document the increase in khat consumption in the two decades since the mid-1970s. Anecdotal evidence and impressionistic observations nevertheless are plentiful. One only needs to look at the proliferation of khat stalls and shops in urban areas, suggesting the expansion of khat markets in both size and number. Increased consumption is also seen in production. Many new plantations of young khat can be seen throughout Harerge. Visitors who have once been to Harerge no doubt will observe that more of the region's prime farmland is gradually being replaced by khat plantations. Another marked indication is the fact that khat is being shipped by airplane and by mail to the Somali and Ethiopian immigrant communities in the Middle East, Europe and North America. Anyone who has seen the fleets of pick-up trucks speeding down the roads from the town of Asba Tafari eastward and from AwWoday westward to Dire Dawa at all hours of the night will appreciate that such traffic represents the pulse beat of the agrarian economy of Harerge.

By the late 1990s khat-chewing had become such a widespread and familiar practice in Harerge and elsewhere in Ethiopia's urban centers that it no longer attracted anyone's curiosity. Attitudes have changed so much that users chew the leaves publicly and without the stigma previously associated with it. This is not to deny that there is still a vocal opposition, but the fact that chewing has become so mundane a practice perhaps might lead to its reluctant acceptance among a broader section of society. In contemporary Ethiopia, there is no religious divide, class distinction or ethnic affiliation when it comes to khat-chewing. It has become the great equalizer and homogenizer.

Botany, chemistry & pharmacology

The khat shrub is a member of genus *Catha* in the family *Celastraceae*, which includes climbing trees and shrubs, with or without spines, that are found widely dispersed in all continents except the polar zones. The origin and nature of the plant have long been subjects of controversy and uncertainty in botanical, medical, pharmacological and political circles. Until recently, many scholars underestimated or completely neglected the existence of distinct varieties of khat, despite the fact that geographical and morphological differences in khat varieties were reported by Forsskal in his *Flora Aegypto-Arabica* as early as the eighteenth century.[24] The neglect is perhaps related to khat's lack of an identified pharmaceutical product or to many researchers' uncritical embrace of a monotypic view of khat. Even Rodinson, despite acknowledging that the Arabic form 'qat' etymologically is older than the Amharic form 'chat' (for *c* developed from *q* and not vice versa) continued to enunciate that *Catha edulis* could be of only Ethiopian origin.[25]

The existence of another species within the genus *Catha*, first reported

[24] Raman Reveri, *Catha edulis Forssk.: Geographical Dispersal, Botanical, Ecological and Agronomical Aspects with Special Reference to Yemen Arab Republic* (Göttingen, 1983), pp. 30–5; Kennedy, *The Flower of Paradise*, pp. 176–7.

[25] Maxime Rodinson, 'Esquisse d'une monographie du qat,' *Journal Asiatique*, 265 (1977), pp. 71–96.

by Forsskal in 1775, has been corroborated by Raman Reveri, whose work threw into doubt the hitherto standard theory among botanists that *Catha* is monotypic. Reveri's study on cytogenetics revealed that the genus *Catha* has two species: *Catha edulis*, which is the only cultivated species at higher elevations in East and Southeast Africa and Yemen, and *Catha spinosa*, which is observed only in the wild state and at lower elevations in Yemen in warm, arid conditions. On the basis of a cytological study conducted at the botanical laboratory at Hohenheim University in Germany and following Darwin's theory of descent, Reveri further argued that *Catha spinosa* is probably the ancestor of *Catha edulis*. His conclusion rests on DNA analysis showing that *Catha spinosa* is a diploid, that is, it can be propagated from seeds, while *Catha edulis* is a triploid, whose seeds are sterile and which therefore must be vegetatively propagated. This means the triploid form of khat has been preserved over the long evolutionary history of its cultivation, an indication that *Catha edulis* evolved from the ancestral *spinosa*.[26]

All cultivated khat is derived from *Catha edulis*, and all discussions of the plant should be confined to this species. A woody shrub, khat can grow up to seventy-five feet (over 30 metres) in height, depending on conditions. In cultivation, however, it is usually pruned to about eight feet (2.5 metres) to facilitate efficient harvesting. Branches are rounded, or compressed in young shoots without spines and hairs. The bole is straight and slender and may be as much as two feet in circumference. The leaves are coriaceous, that is, opposite on flowering shoots and alternate on older ones, with petioles up to 10 cm long and 2 to 5 cm wide, the median veins and the stalks being reddish-brown and becoming greenish-yellow when fully grown.[27]

Khat shrubs thrive best at elevations ranging from 5,000 to 6,500 feet above sea level, with an average temperature between 65°F and 85°F, good drainage and freedom from frost. Interestingly, this ideal khat ecology is also the optimum condition for the cultivation of coffee. Khat, however, is less susceptible to disease and insects than coffee and other highland crops. Although khat shrubs are extraordinarily adapted to highland ecologies, intense and continuous harvest could kill the plant in the long run. The cultivation and harvesting methods of khat have remained relatively unchanged over the centuries.

Seedlings can be planted year-round, but khat is normally propagated by cuttings, most of which are cultivated between January and March. It

26 Reveri, *Catha edulis*, pp. 30–5; Kennedy, *The Flower of Paradise*, p. 177. New research on khat's botany indicates the existence of yet another species, the *Catha transvaalensis*, which is endemic to the Lyndenberg district in Transvaal, South Africa. There is some disagreement as to the taxonomic position of this species, and its inclusion in *Catha* is somewhat uncertain. Sebsebe Demisew, 'Botanical Aspects of "Khat": *Catha edulis* (Celastraceae),' in Ermias Dagne (ed.), *Proceedings of the International Symposium on Khat (Catha edulis): Chemical and Ethnopharmacological Aspects* (Addis Ababa, 1984), pp. 8–9.

27 W.J. Darby *et al.*, 'Khat (or Chat),' in *Ethiopian Nutrition Survey. A Survey by the Interdepartmental Committee on Nutrition for National Defense* (Washington, D.C., 1959), p. 166; Michel Trellu, 'The Pharmacodynamics of Khat,' *Bulletin on Narcotics* 11:2 (1959), p. 43.

requires a good deal of care, with daily watering and a certain amount of manure. A properly cultivated plant will yield its first crop within two to three years and will remain productive for up to seventy-five years. Three to four annual crops are common, but the scant use of fertilizers, extensive cultivation techniques and the absence of adequate irrigation may hamper yield significantly. To minimize damage, all harvesting is done by hand, with each branch picked individually. Leaves close to the ground and those falling during harvest are later packed separately and sold as an inferior-quality product.[28]

In Ethiopia, chewers identify several types of khat, although, based on cultural practices, farmers distinguish essentially two basic cultivars: *deema*, medium-sized reddish leaves and *daalota*, small, pale, yellowish leaves.[29] *Daalota* is less acidic in taste and is preferred by local chewers for its greater effectiveness. The *deema* variety is said to be sturdier and more lasting than the *daalota* and, as a result, came to be exported in greater quantities. Khat is further classified in several ways under a variety of names. But it must be noted that references to varietal differences may not have any relevance for botany since the kinds of khat used are related to cultural practices and to ecological variations; or they signify quality and marketing brands.

The long, tender leaves that bud after the previous year's harvest are designated as *urata* and may be harvested at any time of the year. The young tender shoots are classed as *urata-qartii*, and the slender, leathery ones as *urata-quda*. If for some reason the *urata* is left unharvested for more than six months, the new shoots grow to a new market brand known as *haddaara*, *Hafaa* or *Ijanaqaa*. If harvested while its leaves are young and tender, *haddaara* khat is referred to as *haddaara-qartii*, whereas late-harvested, rough, leathery leaves are known as *haddaara-quda*. Leaves that have grown to full maturity are called *qartii*. Sometimes a leafhopper called '*quda*' feeds on the khat leaves at the *qarti* stage and retards the maturing process. The farmer stimulates new growth by manuring, weeding and shoring up. The renewed leaves are called *quda*, which are considered of high quality and therefore have a very high market value. A poor-quality khat with aged leaves is left on the tree so that women may harvest and retail it in local markets. This type is referred to as *faaqaa*, *taacharo* or *chabala*. If the rains stop before the young shoots, the *urata*, mature, growth is retarded and the plants start to die out. If, however, the rains return or irrigation is made possible, the plant resumes growth, but no longer is considered to provide *urata*, but *kara-bula*.[30]

[28] Amare Getahun and Abraham Krikorian, 'Chat: Coffee's Rival from Harar, Ethiopia, I. Botany, Cultivation and Use.' *Economic Botany*, 27:4 (1973), pp. 356–7.

[29] There are two more varieties, *goba* and *addi*, which, because of their insignificant market value and popularity, are not classified as major types.

[30] My description is based on extensive interviews with informants in Harer. Informants: Ahmed Bakar, Haramaya, May 5, 20 and 27, 1994; Abdurahman Emer, Bate, May 10, 1994; Abdulahi Ahmed, Ganda Issa, May 17, 1994; Mohammed Yuyu, Ganda Issa, May 11, 1994. Cf. Amare and Krikorian, 'Chat: Coffee's Rival from Harar,' pp. 361–3. P.C.M. Jensen, *Spices, Condiments and Medicinal Plants in Ethiopia: Their Taxonomy and Agricultural Significance* (Wageningen, 1981), pp. 161–5.

Photo 1.3 *Khat trees* (daalota *variety)*

Photo 1.4 *Khat leaves* (deema *variety)*

Over the last century most of the research on khat has focused on its chemical constituents, particularly on the isolation of the pharmacologically active compounds considered to be responsible for its stimulating effects. Since early in the twentieth century, owing to transportation difficulties from the Red Sea area to Europe, all experiments were performed on dried leaves of varying quality. The results have been somewhat confusing and contradictory, but most agree that khat's stimulating ingredient was an alkaloid named cathine or *d*-norpseudoephedrine, first isolated in 1913 and identified in dried khat in 1930.[31] Although in subsequent studies it was repeatedly stated that cathine was the main constituent present in the plant, several authors were able to demonstrate the presence of other alkaloidal compounds with a more powerful stimulating action.[32] Science was unable to move on until the 1970s, when faster transportation, new technologies of on-site extraction and the perfection of freeze-drying methods enabled researchers to work with plant material in which the more volatile ingredients were preserved. Studies conducted on fresh leaves subsequently revealed the existence of a highly unstable component called cathinone or (−)-alpha-aminopropiophenone.[33] Tests on animals and observations have now revealed that the new compound is much more important than cathine in producing the stimulant-like activities of khat. It has also been shown that cathinone is structurally and pharmacologically similar to and produces behavioral effects analogous to both norpseudoephedrine and *d*-amphetamine.[34]

There is no agreement or scientific consensus on whether khat is addictive or not. The pharmacological, chemical and medical findings are inconclusive and contradictory. It is nevertheless fairly well established that cathine and cathinone induce temporary euphoria, enhance wakefulness, suppress hunger and increase physical energy and concentration. In the past, some scientists and health professionals believed that the active substances were addictive drugs which, if used over a long period of time, could lead to massive deterioration in the health of the

[31] R. Stockman, 'The Pharmacological Action of *Catha edulis* and its Alkaloids,' *Journal of Pharmacology and Experimental Therapeutics*, 35 (1913), p. 251; G.A. Alles, M.D. Fairchild and M. Jensen, 'Chemical Pharmacology of *Catha edulis*,' *Journal of Medical Pharmacology and Chemistry*, 3 (1961), p. 323.

[32] R. Paris and H. Moyse, 'Abyssinian Tea (*Catha edulis* Forssk., Celastraceae),' *Bulletin on Narcotics*, 10:2 (1958), p. 31.

[33] Kalman Szendrei, 'The Chemistry of Khat,' *Bulletin on Narcotics, Special Issue Devoted to Catha edulis (Khat)*, 12:3 (1980), pp. 5–36; idem, 'Recent Progress in Khat Chemistry,' in Ermias Dagne (ed.), *Proceedings of the International Symposium on Khat (Catha edulis): Chemical and Ethnopharmacological Aspects* (Addis Ababa, 1984), pp. 13–32.

[34] Peter Kalix, 'The Pharmacology of Khat and of the Khat Alkaloid Cathinone,' in B. Shahandeh, R. Geadah, A. Toungue, E. Toungue and J. Rolli (eds), *The Health and Socio-economic Effects of Khat Use: Proceedings of an International Conference on Khat* (Lausanne, 1983), pp. 140–3; idem, 'The Pharmacology of Khat,' in Ermias Dagne (ed.), *Proceedings of the International Symposium on Khat (Catha edulis): Chemical and Ethnopharmacological Aspects* (Addis Ababa, 1984), pp. 69–75.

chewer,[35] to madness[36] and even death.[37] Recent researchers do acknowledge that users may become psychologically dependent, but argue that it is more appropriate to speak of habituation than addiction.[38]

There has not been a significant change of opinion concerning the international status of khat since the United Nations Expert Committee on Addiction Producing Drugs recommended in 1957 that the plant need not be classified within the accepted definitions of a narcotic or addiction-producing drug.[39] In regard to the pharmacological effects of khat-chewing, the literature indicates that chronic use in large amounts could result in adverse health effects. It has been reported that the inaugural symptoms of chewing khat vary from unpleasant feelings and dizziness to tachycardia and epigastric pain. The unpleasant feelings of the inaugural phase are gradually replaced by bliss, euphoric and overly energetic pleasant feelings. Long-term use of khat is also reported to result in insomnia, nervousness, impotence and nightmares. Gastrointestinal problems such as anorexia and constipation are common phenomena in long-term khat use. It is well established that anorexia leads to malnutrition and increased susceptibility to infectious diseases, especially tuberculosis. Furthermore, chronic use of khat is known to have reproductive toxicity in human beings and in experimental animals. Several studies indicated significant decreases in testosterone levels, sperm count and motility and increased number of abnormal sperms in cathinone-treated male animals. Reduced birth weight of babies and inhibition of lactation have also been reported in khat-chewing mothers.[40]

Popular perceptions of the effects of khat-chewing diverge from the observations of the scientific community. Experienced chewers suggest that the amount of the chemical in the leaves varies with production area and variety.[41] The effect on users also varies from person to person. One consumer, for instance, suggested to me that he chewed khat not to

[35] Abdulahi S. Elmi, 'The Chewing of Khat in Somalia,' *Journal of Ethnopharmacology*, 8 (1983), p. 165. H. Halbach, 'Medical Aspects of the Chewing of Khat Leaves,' *Bulletin of the World Health Organization*, 47 (1972), pp. 21–9; Dr Kervingant, 'The Consumption of Khat in French Somaliland,' *Bulletin on Narcotics*, 11 (1959), p. 42.

[36] J.C. Carothers. 'Mirra as a Cause of Insanity,' *The East African Medical Journal*, 22 (1945), pp. 4–6. Carothers's views are often described as racist. His characterization of khat (mirra) as a cause of insanity is seen as a reflection of his view rather than as a scientific assessment.

[37] R.B. Heisch, 'A Case of Poisoning by *Catha edulis*,' *The East African Medical Journal*, 22 (1945), pp. 7–9.

[38] UN Commission on Narcotic Drugs 'The Medical and Social Problems of Khat in Djibouti,' *Bulletin on Narcotics*, 9:4 (1957), pp. 34–6; Halbach, 'Medical Aspects of the Chewing of Khat Leaves,' pp. 21–9; Jensen, *Spices*, pp. 156–71.

[39] UN Commission on Narcotic Drugs, 'Catha edulis (Khat): some introductory remarks,' *Bulletin on Narcotics. Special Issue Devoted to Catha edulis (Khat)* 32:3 (1980), pp. 1–3.

[40] WHO Advisory Group, 'Review of the Pharmacology of Khat,' *Bulletin on Narcotics. Special Issue Devoted to Catha edulis (Khat)* 32:3 (1980), pp. 83–94.

[41] In the Harer plateau of eastern Ethiopia, farmers claim that *deema* khat grown in the immediate vicinity of Harer is quicker to induce hilarity than the *daalota* variety grown in the Chercher highlands.

stay awake or to 'get high' but because he enjoys chewing, indicating that many contemporary users are more interested in the social aspect of chewing with friends than in the religious dimension or the simple experience of getting high. Others claim that, contrary to the general perception that khat is a sleep repellent and appetite suppressant, they have gone to sleep or have felt hungry and thirsty while chewing the leaves. Scientists have a long way to go before they are able to convince traditional users that khat is harmful to their overall well-being.

Khat, agriculture & Ethiopian historiography

Although agriculture and farmers comprise approximately 75 percent of Ethiopia's economy and population, smallholder agriculture has not been given the scholarly attention its importance may warrant. While *Leaf of Allah* should be read primarily as a book on the history of khat from the late nineteenth century to the early 1990s, it is also meant to be read as an interpretation of broader issues of agrarian transformation. Khat is used as a unifying concept to explore demographic and ecological changes, the role of unofficial cross-border commerce and the effects of interventionist state policies on agriculture. The approach taken in this book is to view khat as part of the cropping system in Harerge rather than as a simple psychoactive substance illicitly grown for cash. The central idea of the study is to analyze farmer agency – the decisions farmers make, individually and in the aggregate, their management of changing social and political situations and agricultural contexts. As a study of smallholder agriculture, this book focuses, in part, on farmers' seasonal, annual and long-term choices, agronomic practices and farm management strategies. Following James McCann, I prefer to view the actors here as farmers, but, in this chapter alone, I refer to rural producers as peasants since that is the term generally used in the literature on Ethiopian agriculture.[42]

The literature has generally focused on such themes as the impact of government policies on agricultural production, the failure of successive Ethiopian governments to stimulate the agricultural sector, the effects of Ethiopian land tenure systems on agricultural production, and the exploitation of the tiller by landowners. In the pre-1974 period, scholarly studies generally emphasized the impact of government-sponsored experiments on commercial agriculture. After the revolution of 1974, the focus largely shifted to measuring the extent to which the land reform proclamation of 1975 had succeeded (or failed) in improving agricultural production and ameliorating the condition of the peasantry. In these assessments, peasants, the supposedly active players of the sector, remained

[42] James McCann, *People of the Plow: An Agricultural History of Ethiopia, 1800–1990* (Madison, 1995), pp. 6–7. For a review of the literature on African peasantry, see Allen F. Isaacman, 'Peasants and Rural Social Protest in Africa,' *African Studies Review*, 33:2 (1990), pp. 1–120.

largely inactive and invisible, depicted as victims of landlords and the environment, incapable of self-improvement without government support and subsidies.

The literature emphasizes that Ethiopia's land tenure system hampered peasant production, since either the government or the landowners, or both, heavily taxed produce. The imperial government (1941–74) placed little importance on peasant agriculture, favoring instead the development of the physical infrastructure and commercial or mechanized agriculture. The process of modernization led to peasants being evicted from their farm lands and becoming landless rural laborers.

The *derg* confiscated land and redistributed it among peasants, but failed to provide the tools and materials necessary to increase production. Its policies of direct involvement in the marketing of agricultural products provided further disincentives, leading to declining production. In the end, the studies concluded that peasants remained subsistence agriculturalists whose main goals were to avoid risk and insecurity.[43]

In portraying Ethiopia's cultivators as subsistence-oriented, most students of Ethiopian agrarian history have uncritically accepted such venerable notions as peasants produce barely enough for their subsistence, are generally reluctant to produce surpluses for the market or to enter into cash or exchange relations. Peasants are said to operate in a non-capitalist mode of production, the goal of which is neither accumulation nor investment of capital, but simple reproduction, that is, a quest for subsistence and replacement through the intensive use of family labor. Influenced by modernization[44] and neo-Marxist[45] theories, academics and

[43] Some economists have characterized the Ethiopian peasant as pursuing 'farming practices which are very much in the biblical tradition of his forefathers.' Assefa Bequele and Eshetu Chole, *A Profile of the Ethiopian Economy* (Addis Ababa, 1969), p. 112. Even Desalegn Rahmato, the putative expert on Ethiopian peasant agriculture, recently argued that 'the logic of peasant production centered around self-sustenance and the preservation of existing standards of life.' Desalegn Rahmato, 'Peasant Agriculture under the Old Regime, ' in Shiferaw Bekele (ed.), *Economic History of Modern Ethiopia, 1935–1974*, vol. 1 (Oxford, 1995), p. 170.

[44] Conceived in the 1950s, 'modernization theory' divides societies into modern and traditional camps, respectively represented by the advanced capitalist societies and the less fortunate societies in the Third World. The traditional/modern dichotomy is replicated in the rural areas of traditional societies by a sharp distinction between a dynamic, growth-oriented (capitalist) agriculture and a stagnant, subsistence-oriented (peasant) agricultural sector. For details, see Sydel Silverman,'The Peasant Concept in Anthropology,' *The Journal of Peasant Studies*, 7:1 (1979), pp. 49–57; D.C. Tipps, 'Modernization Theory and the Comparative Study of Societies: A Critical Perspective,' *Comparative Studies in Society and History*, 15:2 (1973), pp. 200–9; P.F. Klaren and T.J. Bossert, 'Lost Promise: Explaining Latin American Underdevelopment,' in idem, *Promise of Development: Theories of Change in Latin America* (Boulder, 1986), pp. 9–10.

[45] Prevalent in neo-Marxist studies of peasantries and agrarian change of the 1960s and 1970s was the idea that peasant production was not only subsistence-oriented but also doomed to disappear. This vision was informed by the notion that persistent poverty in developing countries was the result of surplus extraction by industrial capitalist societies. The outward flow of wealth is replicated at different levels (nation-state, rural regions, peasant communities, peasant households), and, at the economic unit or household level, peasants are simply unable to accumulate wealth, for their surpluses are constantly

policy-makers in Ethiopia considered peasant production an obsolete remnant of the past, which needed to be adapted to modern times. Convinced that large-scale mechanized agriculture was the only road to economic development, they advocated that the state gradually compel the peasant population to modernize its agricultural practices. Despite the obvious failure of government-sponsored agricultural ventures in Ethiopia, some aspects of such an approach still persist in some academic and policy-making circles.

This book sustains the idea that, whether in academe or around seats of authority, it is just as dangerous to neglect the logic of peasant production as it is to neglect the place of the peasant producers in society as a whole. Large numbers of Oromo cultivators in Harerge have over the last half-century massively taken up the cultivation of khat as their main cash crop, and that fact alone contradicts the well-entrenched assumption that peasant agriculture is invariably subsistence-oriented. In Harerge, we rather encounter cultivators who responded quickly, decisively and successfully to a range of state policies that threatened their livelihood. Their response centered on vigorously pursuing a strategy of producing a valuable commodity for the market rather than maintaining the traditional crop regime of interplanting food and cash crops. Well aware of market prices and returns and cognizant of the environmental, economic and political risks involved, the cultivators constantly and quite often successfully maneuvered around and undermined state-sponsored avenues of production and exchange. Even though Harerge's farmers were in large measure responding to agricultural development policies that were progressively undermining their economic livelihood, in choosing a very profitable cash crop as their main produce, they also went far beyond the goal of simply satisfying their 'subsistence needs' to accumulating wealth and capital.

Neither were Harerge's farmers the so-called 'risk averters' who placed a high premium on minimizing and avoiding risky economic strategies and maximizing security of tenure and subsistence.[46] The

[45] (cont.) being 'drained away' by capitalists. The neo-Marxists also argue that, because peasants are well aware of the fact that their insertion into capitalist exchange and production relations is ultimately detrimental to their livelihood, they manifest a deeply rooted 'aversion to' capitalism and strive to remain outside the capitalist market. For details on neo-Marxist literature on peasants, see David Lehmann, 'Two Paths of Agrarian Capitalism, or a Critique of Chayanovian Marxism,' *Comparative Studies in Society and History*, 28:4 (1986), p. 603–6; Alain de Janvry, *The Agrarian Question and Reformism in Latin America* (Baltimore and London, 1981), pp. 19–20.

[46] Concerning the risk- and decision-making literature, see F. Cancian, 'Risk and Uncertainty in Agricultural Decision Making,' in P.F. Bartlett (ed.), *Agricultural Decision Making: Anthropological Contributions to Rural Development* (New York, 1980), pp. 161–76; A. Johnson, 'Security and Risk Taking Among Poor Peasants: A Brazilian Case Study,' in George Dalton (ed.), *Studies in Economic Anthropology* (Washington, D.C., 1971), pp. 144–52; Clifton R. Wharton, 'Risk, Uncertainty, and the Subsistence Farmer: Technological Innovation and Resistance to Change in the Context of Survival,' in George Dalton (ed.), *Studies in Economic Anthropology* (Washington, D.C., 1971), pp. 151–77; M. Chibnik, 'Double-Edged Risks and Uncertainties: Choices about Rice Loans in the Peruvian Amazon,' in E. Cashdan (ed.), *Risk and Uncertainty in Tribal and Peasant Economies* (Boulder, 1990), pp. 279–302.

strategies that the khat-cultivating farmers of the Harerge highlands pursued entailed a great deal of risk. They faced, for instance, environmental threats such as drought, volatile markets in the Horn of Africa, escalating ethnic warfare and inadequate transport. The gravest risk to the peasants' livelihood was not even ecological, economic or the volatility of the region's politics, but repressive state policies designed to curb or reverse the spread of khat cultivation and gain control over the cultivation of other cash crops such as coffee. It is in response to these policies that the Oromo farmers of Harerge elected to switch to an alternative cash crop, khat, whose market viability was uncertain and which could not be used for food.

The scholarship on peasant social protest in Ethiopia rightly casts peasants as active defenders of their interests. In some studies, Ethiopia's peasants are depicted as active agents who protest or rebel against high taxes,[47] the burden of sustaining rulers with their retainers and large number of soldiers[48] and the forcible incorporation of their regions into the Ethiopian state.[49] Peasants also resisted government and merchant pressures to promote cash-crop production on an appreciable scale.[50] Indeed, peasants actively defend themselves against intrusive forces, but rebellion is not always their first resort. Peasants calculate their potential gains and losses before they decide to achieve their goals through violent means. Thus, our assessment of peasant protest should move away from the narrow understanding of rebellion as the only means of expressing discontent to showing the whole range of options available to peasants. It is also important to recognize that threats to the peasants' way of life are not always posed by a powerful state apparatus and its personnel, but also by the forces of supply and demand, to which they respond differently. This book seeks to show that peasants have a variety of ways of expressing discontent and defending their interests apart from violent protests.[51]

[47] Donald Crummey, 'Banditry and Resistance: Noble and Peasant in Nineteenth Century Ethiopia,' in Sven Rubenson (ed.), *Proceedings of the Seventh International Conference of Ethiopian Studies* (Addis Ababa, East Lansing and Uppsala, 1984), pp. 263–77; Timothy Fernyhough, 'Social Mobility and Dissident Elites in Northern Ethiopia: The Role of Bandits, 1900–69,' in Donald Crummey (ed.), *Banditry, Rebellion and Social Protest in Africa* (Portsmouth, 1986), pp. 151–72.

[48] Merid Wolde Aregay, 'Millenarian Traditions and Peasant Movements in Ethiopia, 1500–1855,' in Sven Rubenson (ed.), *Proceedings of the Seventh International Conference of Ethiopian Studies* (Addis Ababa, East Lansing and Uppsala, 1984), pp. 257–62; James McCann, 'The Political Economy of Rural Rebellion in Ethiopia: Northern Resistance to Imperial Expansion, 1928–1935,' *International Journal of African Historical Studies* 18:4 (1985), pp. 601–23.

[49] McCann, 'Political Economy'; idem, *From Poverty to Famine in Northeast Ethiopia: A Rural History, 1900–1935* (Philadelphia, 1987).

[50] Charles McClellan, 'State Transformation and Social Reconstitution in Ethiopia: The Allure of the South,' *International Journal of African Historical Studies*, 17:4 (1984), pp. 657–75.

[51] I have drawn on influential interpretations of peasant protest such as James Scott, *The Moral Economy of the Peasant: Rebellion and Subsistence in South East Asia* (New Haven, 1976); Samuel L. Popkin, *The Rational Peasant: The Political Economy of Rural Society in Vietnam*

Khat, traders & conceptualizing smuggling

The expansion of khat production and marketing during the past four decades presents an interesting case study to gauge the adequacy of the analytical models offered to explain the origins and maintenance of trading by unauthorized (unlicensed) partners who do not comply with official marketing regulations. Scholarly interest in the subject is a comparatively recent phenomenon. A few economists published articles in the 1930s, 1940s and 1950s, in which they laid the microeconomic foundations for studying unofficial market structures,[52] but interest in the subject was not sustained. Since the 1970s, however, smuggling in developing nations and trading among legally unauthorized partners and unofficial market structures have been important themes in development literature.[53]

The most debated issue has been the nomenclature of unofficial marketing structures: markets can be parallel, black, fragmented, segmented, underground or subterranean, hidden, second, informal and so forth. Coined by Keith Hart in his 1971 study of informal income opportunities for the urban unemployed in Ghana[54] and popularized by the International Labour Organization in its World Employment Program mission reports,[55] the 'informal sector' became the favorite terminology employed by researchers. Neither the phenomenon nor the concept was new for anthropologists of non-industrial society, who had discovered that goods and services could be made, traded, swapped and bartered among informal networks without involving officially recognized enterprises.[56]

[51] (cont.) (Berkeley and Los Angeles, 1979). For works on Africa, see Robert Bates, *Rural Responses to Industrialization* (New Haven, 1976); Steven Bunker, *Peasants Against the State: The Politics of Market Control in Bugissu, Uganda, 1900–1983* (Champaign-Urbana, 1987); Abner Cohen, *Customs and Politics in Urban Africa: A Study of Hausa Migrants in Yoruba Towns* (Berkeley, 1969).

[52] Kenneth E. Boulding, 'A Note on the Theory of the Black Market,' *Canadian Journal of Economics and Political Science*, 13 (1937), pp. 115–18; Martin Bronfenbrenner, 'Price Control under Imperfect Competition,' *American Economic Review*, 37 (1947), pp. 107–20; Michael Michaely, 'A Geometrical Analysis of Black Market Behavior,' *American Economic Review*, 54 (1954), pp. 627–37.

[53] Edgar Browning and William Patton Culbertson, Jr, 'A Theory of Black Markets under Price Controls: Competition and Monopoly,' *Western Economic Journal*, 12 (1974), pp. 175–89; Jagdish Bhagwati and Bent Hansen, 'A Theoretical Analysis of Smuggling,' *Quarterly Journal of Economics*, 87 (1973), pp. 172–87; Jagdish Bhagwati and T.N. Srinvasan, 'Smuggling and Trade Policy,' *Journal of Public Economics*, 2 (1973), pp. 337–89; Munir A. Sheikh, 'Black Market for Foreign Exchange, Capital Flows and Smuggling,' *Journal of Development Economics*, 3 (1976), pp. 9–26; Mark Pitts, 'Smuggling and Price Disparity,' *Journal of International Economics*, 11 (1981), pp. 447–58.

[54] Keith Hart, 'Informal Income Opportunities and Urban Unemployment in Ghana,' *Journal of Modern African Studies*, 11 (1973), pp. 61–89; D. Mazumdar, 'The Urban Informal Sector,' *World Development*, 4 (1976), pp. 655–79.

[55] For a review of the ILO reports, see Caroline O. Moser, 'Informal Sector or Petty Commodity Production: Dualism or Dependence in Urban Development,' *World Development*, 6:9 (1978), pp. 1041–64.

[56] Marshal Sahlins, *Stone Age Economics* (Chicago, 1972).

Following Hart's work, though, researchers in a number of related disciplines extended the informal sector concept to describe all kinds of actors and a varied range of economic activities that elude the monitoring and extractive capacity of the state.[57] The concept lacked cohesiveness and there appeared at least sixteen other labels that attach to the activities in this area.[58]

The complexity and heterogeneity of informal economic activities are partly the reason why debates over their nature and characteristics have not produced a terminology that could aptly describe the phenomenon. Most of the labels connote illegality, marginality, corruption, extortion and bribery. In most cases, traders are portrayed in antagonistic relation to the state. In fact, when circumstances are conducive, traders who take part in unofficial exchange activities prefer to operate in the 'formal' or 'legal' sector. The state, however, which has the ultimate power to determine what is legal or illegal, keeps moving the dividing line between the two spheres by raising taxes, introducing regulations and banning products. Considering any action taken by governments to increase the enforcement capability of the state as an assault on their independence, traders respond with active counter-strategies rather than being victimized.

The participants in Harerge's contraband trade are involved in the production, processing, transport and distribution of legal commodities. They are ordinary farmers, traders and urban dwellers, who seek to achieve and maintain a decent standard of living. Their activities, at least in the sight of the local society, are by no means intrinsically immoral or criminal. In our case, there is a considerable amount of documentation that attests to traders' desire to maintain their legal standing. Illegality stems from the fact that people who engage in such works ignore administrative codes, both by evading customs authorities and by failing to obtain the licensing necessary to legitimize their work.

[57] For works on the informal sector in the United States and Britain, see Louis A. Ferman, Stuart Henry and Michele Hoyman (Special issue editors), *The Informal Economy: A Special Issue of the Annals of the American Academy of Political and Social Science*, 493 (1987).

[58] Stuart Henry, 'The Working Unemployed: Perspectives on the Informal Economy and Unemployment,' *Sociological Review*, 30 (1982), p. 465. In an attempt to make distinctions between the different terminologies, researchers have suggested at least three categories: the *irregular economy*, the *social economy* (sometimes further divided into a *household economy* and a *communal economy*) and the *underground, hidden or black economy*. For detailed discussions, see L.A. Ferman, L. Berndt and E. Selo, 'Analysis of the Irregular Economy: Cash Flow in the Informal Sector,' Report to the Bureau of Employment and Training, Michigan Department of Labor, Institute of Labor and Industrial Relations, the University of Michigan/Wayne State University (Ann Arbor, 1978); J.I. Gershuny and R.E. Pahl, 'The Future of the Informal Economy,' *New Society*, 3 (1980); S. Burns, *The Household Economy* (Boston, 1977). Other characterizations include: a *bazaar-type economy* by Clifford Geertz, *Peddlers and Princes: Social Change and Economic Modernization in Two Indonesian Towns* (Chicago, 1963); and two juxtaposed systems of production, one derived from capitalist forms of production, the other from the peasant system, by T.G. McGee, 'Peasants in Cities: A Paradox, A Paradox, A Most Ingenious Paradox,' *Human Organization*, 32 (1973).

The concept of the informal economy or sector is applicable to the nature of contraband trade in Harerge in the sense that it refers to traders who evade regulations that govern exchange of goods and services. The term nevertheless seems designed more aptly to characterize particular types of small firms or individuals who work in various markets, legal or illegal, at very low productivity, than to analyze market structures. The concept that most appropriately characterizes the causes, contexts and consequences of Harerge's contraband trade and unofficial market structures is parallel market.

Even though the term parallel market has been difficult to define, there seems to be a consensus about what it refers to. On the basis of a literature survey, David Lindauer has suggested that 'a parallel market is a structure generated in response to government interventions which create a situation of excess supply or demand in a particular product or factor market.'[59] Researchers studying the phenomenon in developing nations use the term parallel market to characterize and conceptualize unofficial market structures.[60] The term may be used appropriately to describe the unofficial commercial activities and market structures in Harerge, but, for the sake of clarity and precision, it might be necessary to discuss the causes and contexts in which parallel markets arise.

For parallel markets to exist, there must be first a situation of excess supply or demand in the creation of which the state plays a major role. Lindauer emphasizes that initial government interventions in the marketplace and not the body of legal sanctions affecting the unofficial market are the core determinants of the parallel market. A classic example is the imposition of a control regime aimed at limiting the extent of a rise in price when there is a supply shortage. The market does not clear at the control price and both producers and consumers seek alternative channels of trade that would eliminate excess demand at higher prices. Other examples include overvaluation of national currencies, which generates a gap in official versus black market exchange rates, and tariffs and quotas, which influence the selling prices of identical tradable goods that are legal or smuggled.[61]

Second, while the role of government is an important condition, more is involved for a parallel market to exist. The form of government intervention determines the existence of parallel market transactions and structures. Taxes and regulations are two major forms of government intervention that can generate parallel markets. Domestic taxes and import and export duties create a situation favorable for parallel markets to emerge by generating excess demand for commodities at illegal, pretax prices. Regulations (price controls, import quotas, forced sale of

[59] David Lindauer, 'Parallel, Fragmented, or Black? Defining Market Structure in Developing Economies,' *World Development*, 17:12 (1989), p. 1873.

[60] Jean-Paul Azam and Timothy Besley, 'General Equilibrium with Parallel Markets for Goods and Foreign Exchange: Theory and Applications to Ghana,' *World Development*, 17:12 (1989), pp. 1921–30; Michael L. Morris, 'Official and Parallel Cereal Markets in Senegal: Empirical Evidence,' *World Development*, 17:12 (1989), pp. 1895–906.

[61] Lindauer, 'Defining Market Structure,' p. 1874.

commodities to marketing boards, and so forth) also can be responsible for the creation of excess demand at official prices. Prohibitions give rise to illegal markets since they prevent the existence of official markets, but they do not necessarily create market parallelism. This interpretation suggests a distinction between an underground economy, which includes all illegal economic activities, and parallel markets, which may include illegal trade in legal goods.[62]

Here it is important to make a distinction between parallel and 'black,' the term most frequently used to describe illegal economic activity. 'Black' covers markets for prohibited goods such as narcotics for which no legal market exists, and hence the term is not merely a synonym for 'parallel.' Experts distinguish between *black goods*, of which possession is illegal, and *black market*, constituting illegal trade in goods that are themselves legal. The issue of khat smuggling and hidden importation of consumer goods, which we will consider in this book, is about a black (parallel) market, not black goods.

The legality of parallel markets is further complicated by the fact that evasion of taxes and duties leads to an accumulation of 'black money,' that is, 'undeclared or illegally acquired incomes held in the form of currency and other readily convertible assets.'[63] Earned as a result of government prohibitions, such incomes can lead to conditions of excess supply in foreign exchange; if the proceeds of illegal transactions must be 'laundered,' a parallel market that is derived from 'black money' may emerge. In principle, markets in which buyers and sellers violate the 'letter of the law' are illegal. However, parallel markets that emerge to vent excess supply or demand generated by government actions should not necessarily imply that unofficial actions are illegal. Enforcement of the law and expected penalties are important determinants of illegality, since, in some situations, governments may overlook the law and permit those markets to operate freely.

Given the casual and unanalytical manner in which the terms 'black' and 'underground' markets are employed, a clear demarcation between them and parallel markets is necessary. The smuggling out of primary products and the smuggling in of consumer commodities in Harerge are more closely related to the descriptions of parallel markets than the myriad characterizations that have emerged in the literature on legally unauthorized transactions and structures. What makes Harerge's smuggling activities parallel is that they were initiated by the same factors analysts described as conditions necessary for the emergence of parallel markets: excess demand or supply initiated by government actions; government intervention characterized by micro-level policies (taxes, regulations, prohibitions and bureaucratic corruption); policies that lead to the emergence of multiple markets; and buyers and sellers

[62] J. J. Thomas, *Informal Economic Activity* (Ann Arbor, 1992), p. 259.

[63] S. Archaya and S. Madhur, 'Informal Credit Markets and Black Money: Do They Frustrate Monetary Policy?,' *Economic and Political Weekly*, 18:41 (October 1983), pp. 1751–6.

exchanging goods that are legal. Within this framework, this study will demonstrate the complex relationship between contraband and the khat trade. To escape the theoretical dead-end created by the confusion over terminologies, the term 'contraband trade' will be used interchangeably with smuggling to describe unauthorized cross-border transactions while the term 'parallel market' is consistently used to refer to structures that make smuggling possible. The term 'parallel market' is preferred because it denotes that Harerge's unofficial transactions are illegal in the sense that the actors fail to comply with government regulations but deal in goods that are legal.

Scope & organization of the book

Despite the warnings and attempts at limiting its use, khat consumption steadily rose over the last century and more rapidly during the four decades after 1940. This may have resulted from a larger cultural transformation that has taken place in Harerge in a relatively short period of time. A study of the changing cultural meanings that khat may have acquired could produce a fascinating social history, but the purpose of this study is not to present a narrative of the social life of khat as a commodity. The main concern of this book is to explain how people in Harerge became involved with the khat plant and how the relationships between them and the plant have changed over time. Put another way, the book concentrates on people, not on a thing, and aims to reconstruct, from a commodity perspective, the history of the people who cultivate and market khat. By focusing on the production, exchange and expansion of khat consumption, this study will trace and document khat's commoditization and the transformation that the process has engendered in Harerge's agriculture and the lives of the people.

The organization of this book corresponds to a thematic division designed to develop the overall theme. In Chapter 2 I present a brief overview of the physical, historical and human landscape of the Harerge region. The following chapter presents an analysis of the effects of government land policies and improved transport infrastructure (the Djibouti–Addis Ababa railway) on agricultural production and marketing, and highlights the spread of khat production, marketing and consumption before the Italian invasion of Ethiopia in 1936.

In Chapter 4 I analyze the different, yet intersecting, processes of demographic, socioeconomic and political changes that set in motion the expansion of khat production and the market in the 1940s. Moving on to Chapter 5, I will discuss how infrastructural developments facilitated transportation, thereby leading to an expansion of khat consumption in northern Somalia, Djibouti and Harerge's lowlands. I try to document, in this chapter, the reactions of governments to the expansion of khat consumption and steps taken to control its social repercussions. In Chapter 6 I take up the question of how key political and economic events as well as how government decisions led farmers and traders to devise novel marketing strategies. Attempts will be made to show that the expansion of khat cultivation was closely intertwined with an increase in cross-border trade with neighboring countries.

Thereafter, in Chapter 7, I turn to the domestic scene and examine how the expansion of consumption led to the emergence of new marketing structures and how farmers responded to market difficulties experienced by the export sector. In Chapter 8 I examine the expansion of khat production, marketing and cultivation in the context of how state development policies, perceived as inimical by farmers and traders, led to the expansion of Harerge's alternative khat market. The first part of the chapter documents the fact that the expansion of khat cultivation was paralleled by a dramatic increase in contraband trade and, more generally, a decline in coffee and food crop production in areas experiencing severe land shortages. The second part examines peasant strategies and decision-making processes in the face of an attempt by the government to monopolize agricultural production and exchange systems in order to dictate prices and control exports.

In Chapter 9 I will turn to the interpretive aspect of the theme of parallel markets. The major aim of this chapter is to demonstrate that Harerge remained insulated from the problems of agricultural production and marketing that had bedeviled other parts of Ethiopia largely through the flowering of unofficial trade and parallel marketing structures. The role of khat in the interaction and interdependence of Harerge's agricultural highlands and the pastoral lowlands will also be evaluated. The book ends, in Chapter 10, with an analysis of some of the long-term implications of khat's growing importance for Harerge's society and economy.

Two

The Setting

Regional geography & human landscape

Traveling by car in a southeasterly direction from Addis Ababa, Ethiopia's capital, toward the ancient city of Harer, one passes through the country's three major geological formations. The traveler begins the journey on the Northwestern highlands, where the capital is located, gradually descends to the hot and dry Rift Valley system, which takes several hours to cross, and then climbs the Southeastern highlands. The last section rises abruptly in the Bale province east of the Rift Valley lakes, extends north through the Arsi province and eastward through the Chercher and Harer districts, before finally descending to the Issa steppes near the town of Jijjiga. The Hararge highlands form a distinctive topographic feature in the middle and eastern section of the Southeastern highland system. The north-facing slope of the uplands drops precipitously into the Awash valley and then into the hot, dry plains of the Southern Afar Depression. The south-facing slope descends to the Shebbelle plains, which lie at 4,500 to 5,500 feet above sea level, and then gradually slopes to Somalia and the Indian Ocean. The research site is located in the Hararge highlands, which may be divided into three major geographic regions: the Chercher or Western highlands, the Gara Mula'ata massif and the Harer or Eastern highlands.[1]

Lying for the most part between approximately 6,500 and 8,000 feet, the Chercher highlands are characterized by a complex surface configuration consisting of deep valleys and higher plateaux on both the north- and south-facing escarpments. From the northern escarpment, a few perennial streams flow through steep, narrow courses into the Awash basin and the southern plains of the Afar Depression. The

[1] Unless indicated otherwise, my description of the physical geography of the Hararge region is to a large extent based on two sources: Clarke Brooke, 'Settlements of the Eastern Galla, Hararge Province, Ethiopia,' Ph.D. dissertation, Department of Geography, University of Nebraska, 1956; and Emilio Scarin, *Hararino: Richerche e Studi Geografici* (Florence, 1942).

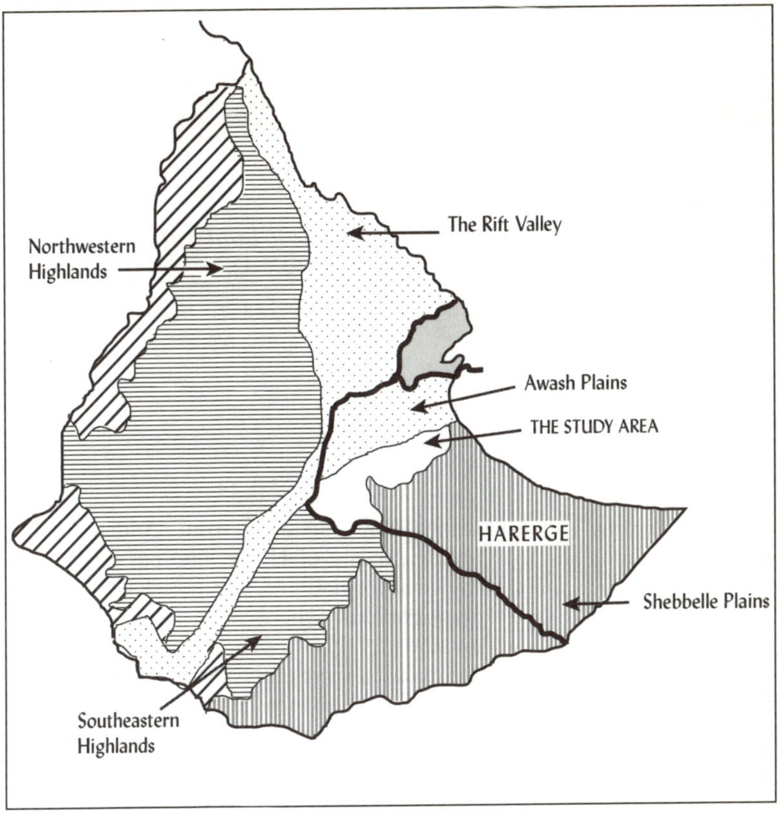

Map 2.1 *Locational map: Harerge in Ethiopia*

volume of water increases during heavy summer rains, causing the streams to carry down the precipitous gorges a great load of alluvial material, which they deposit on flood plains. Along the south-facing slope, drainage into the Shebbelle plains from the high plateau is performed by a number of tributaries to the major rivers such as the Wabi Shebbelle and Genale Dorya.

Many of the south-flowing streams are perennial in character owing to the greater precipitation along the highland's southern slopes and the fact that many of the valleys on the south side tend to hold moisture throughout the year. In high broad basins and alpine meadows of the highlands, water is collected and retained during the rainy season in shallow ponds called *raré*. The water level drops soon after the rains cease, but enough moisture is retained through the dry months to maintain pasture for animals and to water dry-season crops as well as tree crops, especially khat.

At the eastern end of the Chercher highlands, a long spur that protrudes south for a distance of about forty miles forms the Gara

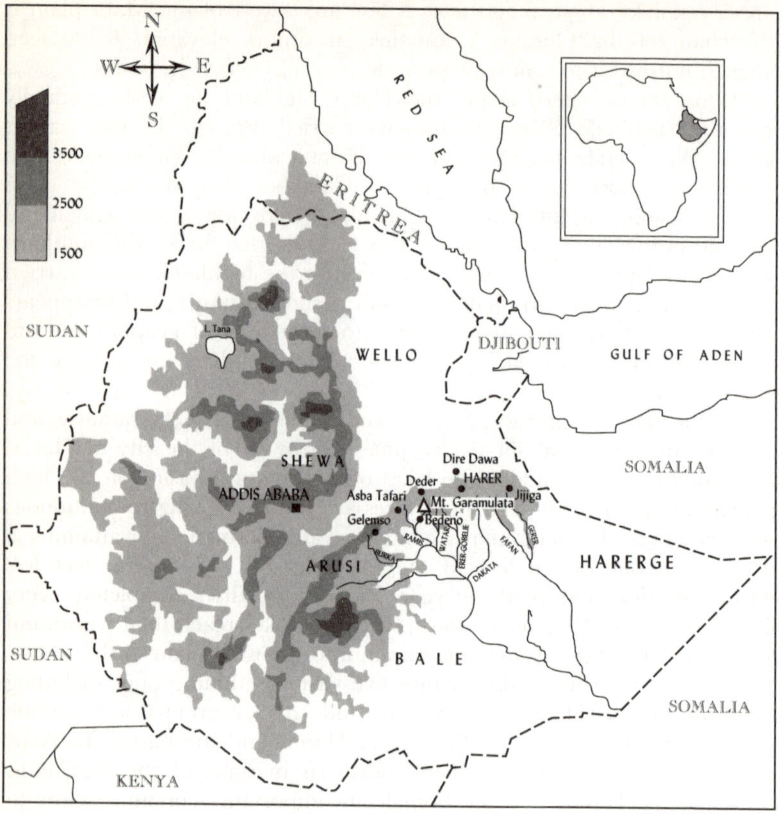

Map 2.2 *Physical map of Ethiopia*

Mula'ata massif. The projected land mass is separated from the surrounding areas by two deep valleys formed by two perennial rivers, the Watar in the east and the Ramis in the west. From the flat tops of the higher mountain flanking slopes drop steeply for about 1,500 to 2,000 feet, but break sharply into broad terraces at elevations about 8,000 to 6,000 feet above sea level, and finally descend abruptly into the valleys on the east and west.

About fifteen miles east of town of Qullubi, the popular pilgrimage site of the shrine of St Gabriel, the Harerge highlands descend to a saddle that narrows to a pass called '*Ya Harer Bar*,' literally the gate of Harer, and then opens to the Harer or eastern highlands, which are characterized by long and terraced valleys of gentle slopes alternating between flat-topped mountains and elongated spurs. The valleys and flat tops are separated in the north from the Afar Depression by the great series of faults that fall sharply for two or three thousand feet into the drier plains below. Bounded in the west by the Gara Mula'ata spur, the Harer highland cascades hills to the Mara plain to the east, which lies at an elevation of about 5,500 feet. A few miles west of the Mara plain is the chain of the Qorahe Mountains, maximum elevation 8,500 feet, aligned northwest to southeast for a distance of twenty-five miles.

Along the southern slope, the Harer highland descends gradually into the Shebbelle plains, its drainage carried through the four valleys of the Watar-Gobellie, the Erer, the Dakata and the Fafan rivers. No perennial streams as in the case of Chercher flow down the steep northern slope of the Harer highlands, except for a few ephemeral natural springs that occur along the foothills that mark the southern wall of the Rift Valley. After heavy rains on the highland, water carried to the foot of the escarpment runs only a short distance and disappears into the lower, warmer plains, owing to the effects of evaporation and rapid infiltration into the sandy cover of the southern margins of the Afar Depression.

In the Harer highlands, two rain-fed lakes, the Haramaya and Addele, are located about twelve miles northwest of the city of Harer. Separated by a low saddle, the lakes occupy natural basins on the high plateau at about 6,300 feet. Both lakes are shallow, owing to continuous sedimentation. The Haramaya contains water permanently, attaining a depth of about twelve feet in the rainy season, but several feet less during the dry months of the year. The Addele dries completely every five or six years. After heavy rains, when the lakes reach their maximum level, water from the Addele may drain into the wide plain to the south.

On these intermountain uplands live various ethnic groups including the Oromo, the Hareri, the Somali and smaller groups such as the Argobba, who live south of the city of Harer, and the pastoralist Afar, who inhabit the largely low-lying northern margins of the highlands. Although the Hareri and the Somali are important economic actors in the region, this study is concerned more with the Oromo, who constitute more than two-thirds of the inhabitants of the Harerge highlands.

The Harerge Oromo,[2] sometimes referred to as the Eastern Oromo, reside in that part of Harerge that stretches from the town of Gelemso in the Chercher to the Qundudo mountain range of the Harer highlands. This area, roughly 200 miles, includes the *awrajas* of Chercher, Wobberra, Habro, Gara Mula'ata, Dire Dawa, Harer, Gursum and Jijiga. The Oromo arrived in this region around 1530–40, when the Barentu, one of the Oromo moieties,[3] migrated east from south-central Ethiopia, drove out the inhabitants of the Harerge highlands toward the coast, and occupied the area south and southeast of the Awash river. The newcomers rejected the political precepts of the Qur'an as well as the suzerainty of the preexisting authorities and established their own

[2] The ethnographic description of the Oromo is based on Atsma Giorgis, *Atsma Giorgis and His Work 'History of the Galla and the Kingdom of Shoa,'* translated and edited by Bairu Tafla (Stuttgart, 1987); G.W.B. Huntingford, *The Galla of Ethiopia: The Kingdom of Kaffa and Janjero* (London, 1955); Demissie, *The Kottu of Ethiopia*.

[3] The second moiety, the Borana, migrated and settled the southwestern and western part of present-day Ethiopia.

political institution, the *gada* system of government,[4] which survived down to the late nineteenth century when it was dismantled by the Egyptians who occupied the region from 1875 to 1885.[5]

The Oromo also formed a complex social organization, which operated on the basis of *gossa,* a patrilineage. Each *gossa* (clan) lived in a village settlement system called *ganda,* which is an agglomeration of 10 to 20 huts comprising 60-100 persons.[6] According to the Oromo's tradition, seven major *gossa* can be identified among the Harer Oromo: Ittu, Anniya, Alla, Baabbille, Obborra, Jarso and Nole. All claim to have descended from a distant common ancestor, Hambana, son of Barentu, through his two sons, Anniya and Qallo. The latter allegedly fathered four sons, Alla, Baabbille, Daga and Obborra. Among these, Daga was the father of three sons, Nole, Jarso and Hume, whose descendants became the most populous Oromo groups in Harerge. Today, the descendants of the four sons of Qallo, known as Afran Qallo, and the other Oromo lineages inhabit the entire Harerge highlands: from the western part of Chercher to the Gara Mula'ata massif, along the northern escarpment of the Danakil Depression, to the most easterly extensions of the Harerge plateau near the headwaters of the Fafan and the Dakata rivers.

Only a century and a half ago, nearly half of the Oromo population on the Harerge highlands were pastoralist. The Hareri, who had possessed a highly developed agricultural economy long before the advent of the Oromo on the Harerge highlands, claim to have taught the Oromo how to farm. Their claim is justified to the extent that the Oromo who settled in the immediate periphery of Harer have borrowed specific crops from the town-dwellers. By the 1840s, only the Baabbille and the mixed Oromo–Somali groups east of the town had adopted the agricultural practice of the Hareri.[7] The majority of the Oromo took up agriculture as their main occupation as a consequence of their participation in exchange relationship with the Hareri who resorted to trade after being expelled from their farm lands by the Oromo.

By the mid-nineteenth century, the Oromo from as far as some thirty miles northeast of the town were fully involved in market exchanges with the town-dwellers, supplying them with much of their basic food requirement and with products that Hareri traders exported. A number of

[4] For a description of the *gada,* see Asmarom Legesse, *Gada: Three Approaches to the Study of an African Society* (New York, 1973).

[5] Muhammad Hassan, 'The Relation between Harar and the Surrounding Oromo between 1800–1887,' B.A. essay, History Department, Haile Sellassie I University (1973), pp. 22–6.

[6] Clarke Brooke, 'The Rural Village in the Ethiopian Highlands,' *The Geographical Review,* 99:1 (1959), pp. 60–5. Demissie, *The Kottu of Ethiopia,* pp. 57–8.

[7] R. Caulk, 'Harar Town and its Neighbors in the Nineteenth Century,' *Journal of African History,* 3 (1977), p. 373. Haberland indicates that the Oromo had possessed a mixed economy of cattle-breeding and the cultivation of barley in their country of origin long before they came into contact with the highland cultivators. Eike Haberland, *Galla Sud-Æthiopiens* (Stuttgart, 1963), pp. 773–4.

Oromo lineages visited the town market with supplies of milk, butter and other products while the Alla from west of Harer furnished coffee, tobacco and safflower, which the town's merchants sent to ports along the Gulf of Aden.[8] Other Oromos hunted elephant in the valleys south of Harer and brought the ivory to the Emir of Harer, who monopolized the trade.[9] In exchange, the Oromo received locally produced or imported cloth as well as salt.[10] The Oromo gradually became incorporated into Harer's economy. The process of their integration was much accelerated during the reign of the Emir Muhammad ibn-Ali (1856–75), who launched a systematic effort to convert the Oromo to Islam and increase their involvement in commercial agriculture. Emir Muhammad encouraged the planting of coffee, sugar-cane, bananas and other commercial crops. The mutual economic dependence that the Hareri and the Oromo around Harer town established at the marketplace gradually led to the transformation of the latter from independent pastoralists to sedentary agriculturalists.

Nevertheless, the Oromo who inhabited the areas beyond thirty miles of Harer remained out of Harer's jurisdiction and influence, governed by their own 'parliament' known as the *Raba–Dori*,[11] which comprised elected political and military officials. Egyptian officials who during their occupation visited the Gara Mula'ata region, about thirty miles southwest of Harer, found that much of the land there was uncultivated and most Oromo were pastoralists. These Oromo, far from taking up agriculture, were exerting pressure on the Hareri and the sedentary Oromo who farmed the land closest to the walls of the town in search of grazing land.[12] Their transformation to sedentary agriculture occurred as the result of the Egyptian administration's consistent effort at increasing agricultural production and their incorporation into the Ethiopian political system in the early twentieth century.

The Hareri, who played such a crucial role in the transformation of the Oromo, spoke a language that belonged to the Semitic family and lived within the walls of the city of Harer. There are several theories

[8] Yusuf Ahmed, 'An Inquiry into Some Aspects of the Economy of Harar and the Records of the Household Economy of the Amirs of Harar, 1825–1875,' *Ethnological Society Bulletin, University College of Addis Ababa*, 10 (1960), pp. 32–6.

[9] The emirs' monopoly of the ivory trade was a longstanding one. In the early nineteenth century, Salt reported that the ivory trade was controlled exclusively by the emirs or their agents. H. Salt, *A Voyage to Abyssinia* (London, 1814), p. lxx.

[10] Hassen, 'Relation between Harar,' p. 1; Caulk, 'Harar Town,' pp. 373–4; Sidney Waldron, 'The Political Economy of Harari–Oromo Relationships, 1559–1874,' *Northeast African Studies*, 6:1–2 (1984), pp. 24–8.

[11] The *Raba* and *Dori* are the political and military wings of the Oromo system of government, respectively. The *Raba–Dori* is essentially the confederated parliament of a number of *gossa*, replicated at each level. For details of the *Raba–Dori*, see Hassan, 'Relation between Harar,' pp. 23–5; G. Douin, *Histoire du Règne Khédive Ismail*, Tome III (Cairo, 1941), pp. 608–9; M. Sabry, *Épisode de la Question d'Afrique: L'Empire Égyptien sous Ismaïl et l'Ingérence Anglo-Française, 1863–1879* (Paris, 1933), p. 427.

[12] Caulk, 'Harer Town,' pp. 381–2. Sabry writes that the Egyptians found it deplorable that so much of the fertile land around Harer was left uncultivated. Sabry, *L'Empire Égyptien*, p. 422.

regarding their origin,[13] but the most likely one postulates that they represent remnants of the Semitic-speaking Harla, an autochthonous population that existed in the region prior to the arrival of the Oromo, practising a highly developed agricultural economy and living in urban centers built of stone. It is probable that, when the Oromo arrived, the Harla escaped to Harer town, which they fortified. There, they were able to preserve their identity for centuries as traders supplying the surrounding Oromo with consumer goods.[14]

By 1900, pursuing trading relations with the Oromo, most Hareri had given up agriculture and taken up trade. Engaged in the import–export trade with the coast, the Hareri traders used two main routes, the Berbera and Zeila routes, to trade with the coastal ports. The latter was preferred by traders because it could be crossed in 15 days by commercial traffic as opposed to 25 days for the Berbera route.[15] Along the Zeila route merchants passed across Issa- and Oromo-populated lands, and goods had to be unloaded from the Issa camels onto Oromo mules and donkeys for final transport to Harer, since camels were not suited to the highland journey.

The Somali comprised an important minority population in the Harerge highlands, but they played the most important role in the transit of Harer's trade. The most easterly of the Somali sub-clans in the highland region are the Gherri-Jarso and the Gherri-Baabbille, who are ethnically mixed Oromo and Somali, practicing agriculture using Oromo farming techniques but possessing Somali social institutions.[16] Further east in the arid lands that extend to the coast of the Gulf of Aden live the Issa. To their south, between Berbera and the foothills of the Jijjiga plain, the land is inhabited by the Gadaboursi.[17] Belonging to the Dir family, both of these clans are camel herders, the Issa in particular occupying a structural niche in the chain of communication and transportation that linked the general region of the Horn to overseas markets. In the nineteenth century, they serviced the export and import trade between the Harerge highlands and Somali ports by organizing a camel caravan led by the *abban*.[18]

[13] Wolf Leslau, *Ethiopians Speak: Studies in Cultural Background, I. Harari* (Berkeley and Los Angeles, 1965), pp. 62–3.

[14] On the Harla, see R.P. Azias and R. Chambard, *Cinque années de recherches archéologiques en Ethiopie* (Paris, 1931), p. 35; Huntingford, *The Galla*, pp. 31, 174.

[15] Peter D. Coats, 'Factors of Intermediacy in Nineteenth-century Africa: The Case of the Issa of the Horn,' in Thomas Labahn (ed.) *Proceedings of the Second International Congress of Somali Studies*, Vol. II (Hamburg, 1984), pp. 180–3. Shiferaw Bekele, 'The Railway, Trade and Politics: A Historical Survey, 1897–1935,' M.A. thesis, History Department, Addis Ababa University, 1982, pp. 1–5. For a detailed description of the caravan journey from Zeila to Harer, see Count Gleichen, *With the Mission to Menelik 1897* (London, 1971), first published in 1898, pp. 5–67.

[16] Scarin, *Hararino*, pp. 33–52.

[17] I.M. Lewis, *A Pastoral Democracy*, 2nd edn (London, 1982), pp. 7–14.

[18] The institution of the *abban*, usually headed by a Somali of the Issa clan at Zeila, was the guarantor or protector responsible for the security of the caravans, the goods and the very life of the trader. It continued to serve the trade between the coast and the

In the last decades of the nineteenth century, the Hareri had not entirely become urbanized traders, but the demographic pressure from the sedentary Oromo and their pastoral kindred was forcing them increasingly to turn to trade. Not all highland Oromo had become sedentary agriculturalists, but those who lived within at least forty miles of the city were tending toward agriculture as the result of the town market. The pastoral Somali were strategically placed to take advantage of the trade between the coast and the interior by monopolizing the means of transportation. By the time of the Egyptian occupation in 1875, the three ethnic groups had worked out a *modus vivendi* whereby the Hareri traded in the agricultural commodities produced by the Oromo by using a means of transportation and communication provided by the Somali.[19]

[18] (cont.) interior until the turn of the twentieth century when some of the traffic was diverted to the Franco-Ethiopian railway. Contemporary contraband traders use the routes and infrastructure of transportation pioneered by the *abban.* I.M. Lewis, 'Trade and Markets in Northern Somaliland,' in P. Bohannan and G. Dalton (eds), *Markets in Africa* (Evanston, Ill., 1965), pp. 365–85; idem, *Pastoral Democracy,* pp. 1–30.

[19] Sebhat Gebre Igziabher, 'Sheik Abadir's City,' *Yakatit* (Amharic weekly magazine, Addis Ababa), 16:1 (1992), pp. 24–5, 36.

Part Two

Khat, Agriculture & Trade

Three

Land & Railway

Structures of agriculture & trade

1875–1935

In Harerge, the spread of khat production as a commodity has not taken place independently of agrarian policies enacted by successive governments. In the eighty or so years before the Italian occupation of Ethiopia, the government of Harerge changed at least four times – from the Emirate (to 1875) to Egyptian (1875–85), to Ethiopian (1887–1935) and, finally, Italian (1936–41). The policies pursued by these governments precipitated a fundamental change in patterns of landholding and in the system of agricultural production and marketing, which, in turn, led to a decisive transformation of the contemporary and subsequent political economy of the region. The arrival of the railway track to Dire Dawa, a town founded at the foothills of the Harerge highlands, was a significant event, which altered the formation of exchange networks and distributional infrastructures for the region's products.

The complex process of agrarian transformation in Harerge that resulted in the commercialization of agriculture, particularly the commoditization of khat, is intimately linked to the change wrought by government policies and market forces in the late nineteenth and early twentieth centuries. On the production side, the landholding systems that evolved as a consequence of government intervention led to land concentration, problems of access for the farmer and eventually land scarcity that influenced the agrarian system. On the marketing side, until the late 1940s, coffee remained a major cash crop dominating farm fields and domestic commerce and the main agricultural commodity that linked the eastern part of Ethiopia to world markets. The amount of cash circulating in the region was subject to the volume of the coffee harvest and the international market prices for the crop. In spite of income fluctuation, the coffee sector was an important source of income for the majority of farmers and for those individuals connected with the selection, processing and packaging of coffee before it could be transported. Beginning in the early decades of the twentieth century, coffee began gradually to lose its dominant status owing to changes in Harerge's system of production and trade that, over time, favored the rise of khat as a dominant cash crop.

Policies affecting agricultural production

In 1875 Egypt conquered the Horn of Africa and won the important inland trading center of Harer. To raise much-needed revenue for their financially beleaguered state, Egyptian officers were as seriously committed to increasing agricultural production as they were to trade. Reforming the land tenure system in Harer was therefore a priority, and arguably their most enduring accomplishment. Essentially following the emirate's landholding system, which overlapped with the administrative structure, the Egyptians appointed officials who were charged with collecting taxes. The appointees were known as the *malaq,* the *garad* and the *damin.* The first two terms were conferred upon officials as a title, but in the course of time they became also appellations for the land under their jurisdiction. Several *garads* and *malaqs,* usually five or six, came under the jurisdiction of the *damin,* literally a guarantor, signifying his responsibility for peace and order among his lineage. The *damin* also served as a liaison between the emir and his subjects.[1]

The Egyptians retained the emirate's land tenure system for tax-collection purposes, but they allowed cultivators to possess land. Prior to the Egyptians, farmers did not have proprietary rights over the land they worked and depended directly on the *garad* or *malaq* for land. In order to ensure the continuous flow of tribute, the Egyptian administration granted cultivators usufructuary rights and the right to transmit land to male heirs or to married females in the absence of male inheritors. The farmer was required to pay taxes to the government based on his harvest, estimated by officials before threshing.

Once land was granted to farmers, the Egyptians took a series of measures to augment production. Land that belonged to the pastoralist Oromo was expropriated and sold to Egyptian officers and powerful Hareri families,[2] who in turn leased it to Oromo farmers. The Egyptians encouraged farmers to plant coffee, as a result of which its cultivation spread from Abadir (the environs of Harer city) to all the districts of the Harer plateau.[3] The Egyptians also introduced to the Oromo new

[1] Yusuf Ahmed, 'An Inquiry,' p. 23.

[2] E. Sylvia Pankhurst, 'Harer Under Egyptian Rule,' *Ethiopia Observer,* 2:2 (1958), p. 56. She refers to Adrien Zervos, *L'Empire d'Ethiopie: le miroir de l'Ethiopie moderne, 1906–1935* (Alexandria, 1936), no page given.

[3] Sabry, *L'Empire Égyptien,* p. 422; Moktar, 'Notes,' p. 382. Coffee trees planted during this time are still found in the districts of Gursum, Gara Mula'ata, Wobberra and Chercher. 'Our Coffee' (Amharic), *Publication of the Ministry of Coffee & Tea Development* (June 1988), pp. 34–5.

vegetables, fruits and cereals, including the grape, potato, beet, almond, peach and the lemon tree.[4]

When King Menelik of Shewa (later emperor of Ethiopia) seized Harer in 1887, therefore, he acquired not only a very rich agricultural region but also a strategically located commercial entrepôt for much of Ethiopia's eastern hinterland. The king's priority was to increase revenue and the state's extractive capacity, but he also had the responsibility of sustaining the soldiers who forced the emir into submission and who stayed on in Harer to enforce the new order. Since the Ethiopian state had no financial capacity to maintain a salaried army and bureaucracy, its fiscal and labor demands could be met by farming households only if they could provide for their own sustenance and the reproduction of the village economy. *Ras* Mekonnen, the first Ethiopian ruler of Harer, fully realized that governmental income and ultimately the power of the state were inextricably linked with the reproduction of the Harerge economy. While the administration had to provide for its soldiers it also had a vested interest in protecting the internal capacity of the conquered communities and a certain minimum of subsistence insurance. The new regime was obliged to adjust its claims to the ability of its tributaries to pay.

The Ethiopian government declared the entire Harerge highlands a Crown Dominion, thus arrogating to itself the legal right to divvy up and distribute land. The Egyptian offices of the *garad* and *malaq* were recognized and office occupants' lands left untouched. Fallow lands and those of the former Hareri ruling families were appropriated by the government and distributed to soldiers, the newly appointed civilian administrators, and the clergy who followed the state in all its ventures to new territories in the south and southwest. Before too long, there was not much vacant land left within a few miles of Harer town to satisfy government needs.

In 1894 *Ras* Mekonnen undertook a series of measures to free up more land for transfer to his large retinue of soldiers and civilian administrators. A land measurement unit known as *qalad* (a rope 75 meters long) or *gasha*, meaning shield, used elsewhere in the empire, was introduced in order to parcel out unmeasured land into measured plots. In the vicinity of Harer, owing to a relatively high population density, one *gasha* out of the surplus holdings of eighty *garads* was confiscated and registered as government domain. In Gara Mula'ata and eastern Chercher, owing to a relatively low population density, one *gasha* was carved out of the holdings of ten *garads* and registered as crown land. To make sure that the indigenous farmers retained reasonable amounts of land, the size of a *qalad* in Harer was reduced compared to the *qalad* size in Shewa.[5] When *qalad* land was

[4] P. Paulitschke, 'Le Harrar sous l'Administration Egyptienne, 1875–1885,' *Bulletin de la Societé Khédiviale de Géographie d'Egypte* 10 (1887). p. 588.

[5] In Shewa, a *qalad* was 7 measures long and 13 measures wide or 9 measures long and 11 measures wide. In Harer, the *qalad* was limited to 6 measures long and 9 measures wide. One measure is 75 meters long.

available, it was distributed among soldiers and civil servants as *maderiya*, in lieu of salary, and to a few voluntary emigrants as *rest*, that is, ownership by reason of settlement or long-term use.

The Ethiopians quickly realized that, given Harer's landscape and a history of independent statehood, cultural assimilation, rather than a protracted military occupation, would be a more expedient way of integrating Harer into Ethiopia's political and socioeconomic fabric. They encouraged people from Shewa and other northern provinces to emigrate and settle among the Hareri and Oromo cultivators, and 'Amhara' settlers arrived in Harer, especially in the wake of the rinderpest-induced famine of 1888–92, which devastated the Ethiopian highlands. No census data exist to show the number of 'Amhara' who settled in Harer, but it is not unreasonable to assume that the number of newcomers was large enough to create a shortage of crown domains.[6]

The scarcity was especially acute in the vicinity of Harer, the area where the highland population was concentrated. *Ras* Tafari Mekonnen (later emperor Haile Sellassie), who was made governor of Hararge in 1910, decided that the endowment of government land could be increased by placing a ceiling on the holdings of the *garad* and *malaq*. Accordingly, an extensive land survey, in which a measurement unit called *jarib*[7] was used, was undertaken in 1912. Land was measured and ceilings were placed on the holdings of the *garads* and *malaqs* at 600 (77.52 ha) and 300 *jaribs* (38.76 ha) respectively.[8] Most of the surplus land that resulted from the measurement was designated 'excess land of the *garad*,' and was registered as government land. Some portion of the land acquired in this manner was sold to original occupiers who had the wherewithal while the rest was divided up and apportioned to new settlers, assigned to military personnel and government officials as *maderiya* or donated to notable chiefs and favorites as *rest* or permanent holding.[9]

In 1919 the government sought to increase the number of tax-paying tenants and, in order to attract new settlers, it reduced the rates of land tax set by the Egyptians. New *garads* were appointed and tenants were settled in the sparsely populated Obborra and Metta lands in Chercher. As co-recipients of land along with the *garads*, the new tenants were entitled to usufructuary rights and the right to transfer the property to offspring as long as they paid their land taxes. The heritability of land

[6] Negatu Wolde Giorgis, *Planning and Implementing the Development of Hararge Province* (in Amharic) (Harer, 1960). A 1938 estimate by the Italian Colonial Administration puts the number of Amhara settlers at 300,000. F. Santagata, '*L'Harar: territorio di pace e di civilità* (Milano, 1940), pp. 91–106.

[7] One *jarib* is 60 standard *kend* long by 60 wide which is equal to 1,296 square meters or 36 meters by 36 meters. One *kend* equals 60 cm.

[8] 'About the administration of land,' *Minutes of the General Assembly Debate on the Question of Land Tenure*, 1950.

[9] E. Brotto, *Il regime della Terre nel Governo del Harar* (The Land Tenure System in the Government of Harer), translated by Asmarom Legesse and Zerabrook Aberra (1966), p. 121.

rights had the unintended consequence of diminishing plot size. After the Ethiopian conquest, more and more Oromo rejected Christianity as the faith of their conquerors and embraced Islam.[10] Young men began to take as many as four wives, thus increasing the number of children born to a single father. In some families, there were as many as eighteen children who could claim land on the death of their father. When the family property was divided up, the inheritors received very small plots, owing to the large number of recipients in a family. By the 1930s, scarcity of arable land was clearly a serious problem in the Harer highlands.[11]

In 1923 a newly appointed governor of Chercher named *Fitaurari* Tekle Hawariat established the entire Chercher region as Tafari Mekonnen's *gult* (absolute property right) and had every piece of land measured into *qalad*. While the land size of a *garad* in eastern Chercher ended up being of the same size as the *qalad* of Harer, namely, 9 ropes in length and 6 in width (30.37 ha), the *burqa* system[12] of western Chercher was established at 11 ropes in length and 7 in width (43.30ha). The *Abba Burqas*, the traditional land overseers, were removed, and the tenants were made directly responsible to the government. Because coffee was an important crop in this region, Takla Hawariat raised land-use taxes to more than six times what tenants paid in the Harer highlands, thereby rendering the area an unattractive destination even for land-hungry farmers from the Harer highlands.[13]

When *Ras* Tafari assumed the Ethiopian throne in 1930, he took a series of steps to ease the problem of land shortage in the Harer highlands. In September 1930 he issued an order that required all lands in Harerge to be measured by the *qalad*. In April 1931 the then governor, Dejazmatch Imeru Haile Sellassie, measured all lands in Harerge province, leaving the *garad* and the *malaq* with 600 and 300 *jaribs*, respectively. The remainder was measured by the *qalad* unit, and the surplus was turned over to the government for distribution among the new settlers under the auspices of a *malkegna*, a member of the gentry who was awarded land for his service to the government. The distribution of land was not limited to grants as wages (*maderiya*). Settlers from the rest of the empire were awarded land for service to the government as *metoriya*, in lieu of pension.[14]

The Ethiopians implemented their land policies in the eastern Harer highland regions, where the land was rich and densely inhabited. Whereas they had laid the foundations of early colonial rule on the basis of traditional institutions, the new tribute requirements and the very

[10] Atsma Giorgis, *History of the Galla*, pp. 27–8.

[11] Negatu, *Planning*, p. 17.

[12] Until the 1920s, the *garad* system, mixed with the *qalad* system, predominated in eastern Chercher. In western Chercher, the communal system was transformed into the *burqa* system in which the *Abba Burqa*, meaning father (owner) of land, executed administrative functions similar to those of the *garad* in the Harer and Gara Mula'ata regions.

[13] Negatu, *Planning*, p. 16; Brotto, *Il regime*, p. 28.

[14] Brotto, *Il regime*, pp. 28–30.

circumstances under which tributaries had to meet their new obligations forced the Oromo communities to alter their way of life. The new overlords, the *malkegna*, commandeered farmers for their own enterprises but – unlike the emirs or the Egyptians, who expected to be paid only in labor – the *malkegna* demanded tribute in labor, in kind and increasingly in cash. Alienation was rampant, on account of failure to pay taxes and promises of taking care of aged title-holders and 'friendship,' a dubious arrangement suggesting forced acquisition. In the process, more land came under the possession of the church and powerful aristocrats including *Ras* Tafari, who, in effect, became the landlord with vast tracts of land in Chercher and most of the Harerge highlands.[15]

The land policies of the Ethiopians in Harerge had two implications for agricultural production. First, land became concentrated in the hands of a patrimonial state that had an insatiable appetite for more land. Officials of the Ethiopian state and its supporters came to possess large tracts of land and confined the indigenous cultivators to their lineage lands, which diminished in size as population increased. Second, the governing elite possessed the right of allocation and hence controlled access to all lands. Land-hungry farmers were discouraged from migrating to new lands, thus increasing population concentration on lineage lands and diminishing the size of the land owned and worked by farming families. In the long run, the difficulty of acquiring land led to increasing fragmentation as tenant holdings were broken up as a consequence of expropriation, population growth and inheritance by multiple heirs.

Cultivators initially responded to the problem of land scarcity by employing several techniques, including moving from densely populated *garads* to thinly populated ones. Some tenants settled on the land of *garads*, who admitted them for the customary *bartcha*, a payment in sheep or cattle. Others became sharecroppers working for other tenants who owned large tracts of land. Still others migrated to the area west of Qullubi and east of the Burqa river when the government after 1910 began to create new *garads* in the uncultivated areas. There were also a few instances in which sedentary agricultural Oromo communities from Harer and Gara Mula'ata region migrated to the uninhabited areas of Chercher in the Ramis valley and reverted to pastoralism. These processes, migration, sharecropping and, to a limited extent, reversion to pastoralism, continued well into the 1930s.[16] When these strategies failed to resolve the problem of land shortage, caused by the various policies of the Ethiopians, Oromo farmers began to deal with the problem by

[15] Ibid., pp. 32–40; Tsehai Berhane Sellassie, 'Developments in Land Tenure in Harerge 1887–1935,' in *Social Science Miscellanea 5*, Institute of Ethiopian Studies, Addis Ababa University (n.d.), p. 4.

[16] Mohammed Hassen, 'Menelik's Conquest of Harar, 1887 and its Effect on the Political Organization of the Surrounding Oromo up to 1900,' *Working Papers on Society and History in Imperial Ethiopia: The Southern Periphery from the 1880s to 1974*, edited by D. Donham and Wendy James, African Studies Centre, Cambridge (1987), p. 237.

introducing new farming techniques, including the incorporation of high-value cash crops into the cropping regime.

Trade & the export economy

The Egyptians conquered Harer fully aware of its commercial potential. Once in control, they were intent on developing and exploiting it. At the time of their occupation, however, Harer's long-distance trade was in a depressed state, owing to the lack of security along the trade routes leading to the coast. Reports by European and other travelers confirm that the lack of security and water posed great difficulties for caravans on the way to Harer and back to the coast. A more local problem was the lack of currency that could have been used to replace barter.[17]

During Rauf Pasha's tenure (1875–7), steps to monetize and monopolize commerce in the Egyptian-controlled areas were taken to encourage and increase trade. Rauf Pasha introduced Egyptian piasters to replace the *mahallaq,* the emir's currency, which had no fixed monetary value and therefore did little to discourage barter. Particularly interested in acquiring control of exports, Rauf put the coffee business in the hands of the Egyptian administration. Coffee was to be purchased and sold in cash to facilitate trade and increase the quantity of currency in circulation. The introduction of Egyptian currency not only revived Harer's trade but also helped to draw the region into the world economy through an expanding Egyptian modern market sector.[18]

For security the Egyptians built military posts along the most frequented Harer–Geldessa–Zeila route in order to protect caravans from brigands and dug a number of water wells. The village of Geldessa was almost reconstructed to accommodate more traders, who began to arrive there on a regular basis. With travel between Harer and its outlets made easier and more secure, foreign trade increased dramatically. The market for Harer's products expanded so much that the number of camel caravans jumped from 70 to 400 in the year the Egyptians seized Harer. Imports rose as more and more foreign traders settled in the town and opened retail shops.[19]

The supply of commodities was handicapped by the region's deficient transport system. The caravans were not efficient or fast enough to keep up with rising demands. An important initiative to improve the regional

[17] The Khedive's instruction to Rauf Pasha regarding the occupation of Harer clearly spells out that the motive of the conquest was 'd'établir la sécurité sur le routes qui mènent au Harrar afin de tranquilliser les voyagers et de faciliter le commerce de caravan.' A.A. Reg. 2 *Ordres Supérieurs*, pièce 82, p. 88, O.S. à Rauf Pacha, 16 juillet 1875. Cited in Douin, *Histoire du Règne Khédive Ismaïl*, Tome III, p. 602.

[18] Jonathan Miran, 'L'occupation Egyptienne de Harar, 1875–1885,' Mémoire de Diplôme de Recherche et d'Études Appliquées Est-Africaines et Malgaches, Institut National des Langues et Civilisations Orientales (juin, 1983), pp. 71–2; Sabry, *L'Empire Égyptien*, pp. 421–2.

[19] Paulitschke, 'Le Harrar,' pp. 586–8.

infrastructure took place when construction of a railroad line connecting the Ethiopian hinterland to the French port of Djibouti commenced in 1897. Originally planned to extend to the Blue Nile via Harer and Ethiopia's capital, the railway line was diverted to the Harerge highlands which proved a formidable barrier to which the technology of the time was not equal. The line reached its first major terminal in 1902, a storage site at the foot of the highlands, which became the city of Dire Dawa. As it turned out, rail transport supplied the desired outlets to the sea.

The arrival of a railroad to Dire Dawa led to a fundamental reorganization of the structure and orientation of Harerge's external trade. The volume of export of agricultural products rose significantly compared to Ethiopia's past capacity; the new French port of Djibouti became the principal outlet for Harerge's commerce.[20] The railways created Dire Dawa as the first modern city in Ethiopia whose reason for existence was business. As a local dictum testifies,[21] Dire Dawa, essentially a creation of the Chemin de fer Franco-Ethiopien (CFE), was catapulted to a position of commercial prominence in a short time. From its inception, Dire Dawa played a crucial role in the region's economic activities as a major transit center, a customs-clearing station, the seat of major import–export firms and a major market for the entire Harerge region.

The establishment of Dire Dawa forced a major reappraisal of the structure and orientation of Harerge's trade. Exports through Djibouti began to increase even before the railway reached Dire Dawa. The value of exports registered a substantial increase from just under 700,000 francs in 1900 to almost 7 million francs in 1902.[22] In subsequent years, the exports through Djibouti increased annually, doubling in 1903 and reaching 18 million francs in 1905 and 35 million francs in 1907.[23] The general increase in trade is reflected in specific items. The value of coffee exports increased almost sevenfold, from 210,313 Maria Theresa dollars in 1898 to 1,315,525 MT dollars in 1902. Figures are unavailable for the amount of skin exports prior to the turn of the century, but, in terms of value, skin exports increased almost sixfold from 141,643 MT dollars in 1902 to 772,690 in 1904. After showing a remarkable increase in the last years of the nineteenth century, ivory exports declined gradually, owing to the depletion of animal stocks.

Despite the obvious statistical increase in the trade through Djibouti, the railway did not completely displace the traditional infrastructure of trade. In the initial phase of construction, the railway authorities had to rely on camel caravans to transport rails and sleepers, water, telegraph poles, camping equipment, food, stones, lumber and so forth along the

[20] G. Mackereth, *Economic Conditions in Ethiopia*, Department of Overseas Trade, No. 507 (1932), pp. 13–15; M. Perham, *The Government of Ethiopia*, 2nd edn (London, 1969), p. 178.

[21] *Izih yallaw hullu babur ya tafaw naw*; literally, 'the train spat out everything that is here.'

[22] *Bulletin de Comité d'Afrique* (BCAF) 1901, p. 72; BCAF, 1902, p. 183. The publication generally has no author or title.

[23] BCAF, 1903, p. 192; BCAF, 1906, pp. 79, 167; FO 401/10, Hohler to Grey, May 24, 1907.

construction lines. The Issa, who had earlier provided transportation for the Harer–Zeila trade, took the opportunity to rent their camels to the construction company. Before the railhead reached Dire Dawa, Issa caravaneers, far from being supplanted by the railway, intensified their participation as transporters of Djibouti's incipient trade.[24]

It was everyone's guess that, once the railway reached Dire Dawa, it would become the sole conduit of trade between the Gulf of Aden and Ethiopia's interior. The French undertook the construction of the railway, believing that the resources of Ethiopia's interior would reward them economically and politically.[25] Their activities were geared toward expanding production, principally in the Hararge highlands, and channeling business to the new port. Ever since the opening of the Suez Canal and the rise of Aden as the principal port in the region, the British had dominated the entire regional trade and were wary that the French move might undercut their dominance. In 1899 the British Minister in Ethiopia, John Lane Harrington, warned his government that establishment of the railway at Djibouti would absorb most of Harer's trade and result in serious economic consequences for the British in Aden and Zeila.[26]

In the end, neither the hope of the French nor the fear of the British materialized. A number of circumstances prevented the railway from tapping the 'rich' natural resources of Hararge and channeling the entire trade to Djibouti. By the time the railhead reached Dire Dawa, the company was in financial trouble, and the cargo available for transport was so small that the company had to impose high freight charges. Calculations based on contemporary publications show that there was a significant difference of transportation costs between the traditional caravaneers and the railway. The cost of transporting all kinds of goods between Zeila and Harer was 39.5 MT dollars per ton by camel and donkey, and 44.5 MT dollars if effected by camel all the way. The rail charges between Dire Dawa and Djibouti varied according to the transported goods.[27]

The railway was not cheaper but it provided speed and convenience. Goods were loaded directly from the mules, which had brought them to Dire Dawa from the highlands. Cumbersome goods such as corrugated iron sheets, which were difficult to load on camels or mules, could now be imported without much trouble. However, the company's decision to adopt exceedingly high carriage rates discouraged merchants from using the trains.[28] Traders found little reason to prefer the train, especially for

[24] BCAF, Nov. and Dec. 1899; see also T.L. Gilmour, *Abyssinia, the Ethiopian Railway and the Powers* (London, 1906), p. 17; Hughes Le Roux, *Ménélik et nous* (Paris, 1902), p. 82.

[25] M. Boucoiran, 'La Situation économique de l'Ethiopie,' *Renseignements Coloniaux et Documents* XI (1918), p. 192.

[26] FO 403/298, Harrington to Grey, August 31, 1899.

[27] Richard Pankhurst, *Economic History of Ethiopia, 1800–1935* (Addis Ababa, 1968), p. 325. The calculation is based on a French publication, *Renseignements Coloniaux et Documents* (1906), pp. 193–6 and a British House of Commons Sessional Paper, *Report for the Year 1905–6 on the Trade of Abyssinia* (Cd. 3283–8), 88 (1907), p. 9.

non-perishable items such as coffee and ivory that did not require rapid transport. As a result, caravans originating in Harer continued somewhat unhindered to use the Issa transport service for their trips to Zeila.[29]

The freight charges were high enough to put a damper on the level of export trade, but the situation was worsened by complex customs formalities. Traders and other travelers were required to deal with 'a most irritating bit of officialdom' at Djibouti, Dire Dawa and later Addis Ababa. One traveler described his railway experience:

> Having run the gauntlet [at Djibouti, at Dire Dawa, once again, everything was] turned out of the train, opened, examined, and delayed, … cases once having been opened are not properly fastened again, goods are pilfered from them while they are lying about … and the remnants frequently arrive broken and damaged… . Goods take almost as long to come from Jibouti to Addis Ababa by rail as they did formerly by camel caravan, and the condition in which they arrive is frequently little better.[30]

Such inconveniences had a direct effect on the import–export trade, often increasing the prices of consumer goods in Dire Dawa and Addis Ababa. The high transport costs were further inflated by various charges and customs dues at 12 percent *ad valorem.*[31] Given the very low purchasing power of the people, prices were invariably prohibitive. Excluding traders' profits, prices increased between Djibouti and Addis Ababa by 250 to 300 percent on basic imported items such as the textiles locally referred to as *abujedid.*[32]

The absence of roads connecting production areas with the railway terminal was another factor that stifled the expansion of trade, especially of perishable agricultural products such as khat. There were practically no roads in the Harerge highlands. The cart track originally built by *Ras* Mekonnen in the early 1900s remained impassable throughout the 1920s, all bridges over the numerous ravines and rivers having been washed away. All transport of coffee and grain therefore was by donkeys and to a lesser extent by mules and porters. Camels were rarely used as the climate did not suit them.[33]

Neither the railway company nor the Ethiopian government took steps to rectify the practices that repressed trade along the railroad line. Because the railway was the only modern outlet, the CFE authorities were able to exact high charges on freight carried on their track without fear of competition. In 1925, the company made a sweeping increase of 30 percent in freight rates, thereby causing a dramatic increase in the cost of shipment. Even at the old rates, the rail charge for transporting

[28] R.P. Skinner, *Abyssinia of Today* (London, 1906), pp. 99–101.

[29] FO 403/255, Rodd to Salisbury, June 4, 1897.

[30] C.F. Rey, *Unconquered Abyssinia* (London, 1923), pp. 213–15.

[31] Customs duties at Harer were 8 percent *ad valorem.* This was another incentive for farmers to prefer the Zeila–Harer route to the railway. *Report for the Year 1905–06 on the Trade of Abyssinia* (1907).

[32] J. Duchesne-Fournet, *Mission en Ethiopia, 1901–1903* (Paris, 1908–9), pp. 282–8.

[33] FO 368/1688, 'Notes on the Products of Harer Plateau,' August 1917.

skins from Addis Ababa to Djibouti alone exceeded ocean freight charges for the same amount of skins from Djibouti to New York. Buyers at Djibouti and Aden could not pay more than the price fixed by their European or American buyers, who cared nothing about the increase in freight charges. The amount of skins exported from Ethiopia nevertheless remained steady, not owing to increased use of the railroad but because export cargoes were diverted to other carriers.[34]

Operating under enormously high freight rates, in conjunction with inefficient customs administration and in competition with a relatively cheaper indigenous transport system of caravaneers, the Djibouti–Dire Dawa railway failed to draw much of the anticipated agricultural potential of Ethiopia's interior.[35] Nevertheless, it was not completely inconsequential. Prior to the railway, high-value trade items were preferred for export to make a margin of profit, which otherwise would be consumed by transportation costs in the case of bulky and heavy products. Among the railway's most visible effects was the fact that merchants were able to export bulky merchandise. This situation favored the export of coffee, which had a stable and promising demand.

In the thirty years before the Italian occupation, export trade increased almost ninefold, from 3,367 to 28,497 tons. Coffee was the leading product, especially the 'Hareri' variety, whose exported quantity, until 1921, was ten times higher than the 'Abyssinian.' After 1922, the latter began to show a considerable increase in volume, equaling the former in 1924 and surpassing it in 1927. The amount of hides and skins exports increased even faster than coffee, doubling between 1910 and 1917 and trebling by 1925.[36] From 1922 through the onset of the Great Depression, international coffee prices were favorable and demand stayed high. In response, Hareri coffee exports increased from 4,396 tons in 1922 to 8,578 tons in 1934.[37] Much of the rise in volume was owing to the increase of the share of the 'Abyssinian' variety harvested from wild coffee. The increase in the 'Hareri' variety was so slight that it reflects a decrease in internal consumption rather than increased production.[38]

Khat in the Harerge region to 1935

Before the railway reached Dire Dawa at the beginning of the twentieth century, khat-chewing was evidently rife in Harer, but largely confined

[34] US Department of Commerce, *Ethiopia: Commercial and Economic Survey* (1918), p. 3.

[35] *Report for the Year 1905-06 on the Trade of Abyssinia* (1907), pp. 8–11, 26. Shiferaw, 'The Railway,' pp. 8–16.

[36] Zervos, *L'Empire*, p. 159; Pankhurst, *Economic History*, pp. 335–7, 430–1.

[37] Harold Marcus, 'The Infrastructure of the Italo-Ethiopian Crisis: Haile Sellassie, the Solomonic Empire and the World Economy, 1916–1936,' in Robert Hess (ed.), *Proceedings of the Fifth International Conference of Ethiopian Studies* (Chicago, 1978), p. 559.

[38] US Department of Commerce, *Ethiopia: Commercial and Economic Survey*, Trade Information Bulletin, No. 476 (Washington, D.C., 1927), pp. 7–9.

to the affluent Hareri elite who had surplus income to spend on khat. The price city-dwellers paid for khat was relatively high compared to that for other cash crops, 1 pound of khat leaves selling for one Maria Theresa dollar. In spite of high prices for khat, farmers kept growing food crops and planting coffee.[39] At least three factors explain the continued cultivation of coffee, vegetables and food crops around the city of Harer instead of the apparently lucrative khat. First, the number of consumers who did not grow their own supply of khat was too small to create a proportional rise in demand. Second, the lack of rapid transportation prevented khat from reaching a wider consumer pool. Third, most of the khat trees were owned by absentee farmers who lived within the city wall and maintained gardens around the city for their own use. The surrounding Oromo apparently did not obtain access to khat until the battle of Chelenqo in 1887, when the widows of Hareri men killed in the war, unable to maintain their khat orchards, employed Oromo farmers as tenants to tend their khat fields. The Oromo thereafter distributed the plant among their kin, thus beginning the spread of khat on a larger scale.[40]

The massive conversion of the Oromo to Islam in the late nineteenth century was an important factor in accelerating the spread of khat consumption among them. Oromo informants unanimously report that they acquired the khat plant and the habit of chewing it from the 'Adere,' as they prefer to call the Hareri. This story is consistent with the general phenomenon that the Oromo adopted agriculture from the Hareri. The Oromo, who remained pastoralists and animists until the middle of the nineteenth century, did not have much to do with coffee, vegetables and cereals, let alone with khat. It was only after they became sedentary farmers and accepted Islam that khat-chewing spread among them. In other words, the Oromo started khat cultivation in the process of their shift from pastoralism to sedentary farming, and its consumption spread among them along with their massive conversion to Islam in the late nineteenth and early twentieth centuries. By 1910 khat-chewing had become a widespread practice among the Islamized Oromo, among whom the leaf quickly attained social, cultural and religious importance.

Non-Muslims, however, considered khat-chewing to be a sign of conversion to the Islamic faith. For the Christian 'Amhara' colonists, the consumption of khat was a mark of apostasy. In 1916 one of the accusations leveled against Lij Iyasu, Menelik's designated successor, was alleged indulgence in chewing khat with Muslims during one of his visits

[39] A British Diplomatic and Consular Report, which reported that farmers planted khat for sale in the city, also shows that, in a couple of years before 1900, 150,000 coffee trees were planted in the Chercher district. House of Commons Sessional Papers, *Reports for the Year 1898–1900 on the Trade of Adis Abbaba and Harar*, 92 (1900), p. 83.

[40] Hill, 'Cat,' p. 13.

[41] Hasen Ali, 'A Short Biography of Bejirond Takla Hawariat,' B.A. essay, Department of History, Addis Ababa University (1985), p. 12.

to Harer in the early 1910s.[41] Khat was considered so evil that one 'Amhara' governor of Chercher, *Fitaurari* Takla Hawariat, made eradicating the crop one of his priorities. Immediately after he was appointed Governor-General of Chercher, Takla Hawariat ordered the Oromo to uproot their khat plants and replace them with coffee trees. He also placed a high tax on khat transactions to restrain market growth. Fearing rebellion, his subordinates, the *malkagna*, appealed to *Ras* Tafari, then de facto ruler of Ethiopia, to reverse the decision. Pointing to the apparent debilitating effect of khat on the human body, Takla Hawariat argued that chewing it would lead to reduced productivity, diminished reproductive capacity and increased incidence of mental illness. His efforts at eradication failed nonetheless, since he eventually came to believe that it would be fruitless to carry out the destruction of khat only in Chercher while cultivation of the shrub was permitted in the adjacent Gara Mula'ata and Harer regions.[42]

Takla Hawariat's concern was also shared by a British agricultural expert who visited Harer in 1917. In his report the expert bemoaned khat as a plant whose 'evil influence' must be curbed. His bias aside, the report contained important information confirming that, by the late 1910s, khat had become an integral part of the cropping system of the Harerge highlands. The report's author delves into various botanical characteristics of khat. Propagated from cuttings, khat was not pruned or irrigated, although it received some cultivation. There were different varieties of khat, coarse and fine leaves, which were plucked off for the purpose of consumption. Most of the harvest was said to be consumed locally, but a certain amount was sent to Dire Dawa and thence to Djibouti.[43]

There is no statistical evidence that shows the amount of khat exported from Harer prior to the late 1920s.[44] Oral sources confirm that, until the turn of the twentieth century, fresh khat leaves were sold in local markets and, because of the absence of a speedy means of transportation, only the dried leaves were exported to places far away from the areas of production. It was only after the Addis Ababa–Djibouti railroad reached Dire Dawa that small amounts of fresh khat began to be

[42] 'The Autobiography of Fitaurari Takla Hawariat,' Unpubl. MS Department of History, Addis Ababa University (n.d.), pp. 66–7.

[43] FO 368/1688. Notes on the Products of Harar Plateau, Abyssinia. A Report by Dudley S. Corlett, August 1917. The report was later published under the same title in *The Poona Agricultural College Magazine* (1917), pp. 161–70.

[44] In his 'An Inquiry into the Economy of Harer ..., 1825–1875,' Yusuf Ahmed suggests that 280 camel loads of khat had been exported from Harer to Aden annually. He cites Burton, *First Footsteps*, p. 299, to support his conclusion. In actuality, what was contained in Burton's account was a quotation from an article by James Vaughan, 'Notes on the Drugs Observed at Aden, Arabia,' *Pharmaceutical Journal*, 12:5 (1852). Vaughan's commentary was on the importation of khat into Aden from Yemen, not from Ethiopia. See also Charles Moser, 'The Flower of Paradise: the Part which Khat Plays in the Life of the Yemen Arab,' *Geographical Journal*, 93 (1917), pp. 177–82.

exported by rail to Djibouti, and thence by dhow to Aden.[45] More khat could have been exported from Harer at this time to meet growing demand in the French and British colonies. The potential was not fully realized owing to the absence of a faster means of delivering khat from the production sites to the railhead in Dire Dawa. The story of my informant Adem Aliyu Jaylan, whose grandfather sold khat in the Dire Dawa market in the early 1930s, demonstrates the potential and the limitation on khat export during this time.

> Most of the men from my grandfather's generation were bold. Surprised, we [my informant's generation] asked them why they were uniformly bold. They told us they became bold because they used to carry head-loads of khat to Dire Dawa almost every night. During those years, there was no road leading from here [AwWoday] to Dire Dawa. Neither did anyone come to buy khat as most people do today. My grandfather and his friends carried khat to Dire Dawa on their heads and traveled there on foot. They set off at midnight using lamps to light their way. It is said that there was a pass on the way to Dire Dawa where, if the lamp-carrier was not very careful, a blowing wind blew the lights out. If that happened to anyone, the person usually ended up getting to Dire Dawa late for the early morning market at *Chat Terra*, the khat market. The farmers must arrive at the khat market by 6 a.m. if he were to sell his khat at good price. The khat leaves that our grandfathers sold in Dire Dawa were exported to an Arab country called Aden.[46]

In spite of the limitations, in the late 1920s and early 1930s, the amount of khat exported to Djibouti seemed to have caught the attention of the government so as to appear in the official trade statistics of the French colony. There, the majority of khat-chewers were Arab merchants and employees of the port facilities. Djibouti's Arab population in the mid-1930s numbered more than 4,000.[47] In 1929 the total amount of khat imported into Djibouti was 13 tons, valued at 322,755 francs. In 1931 the amount more than doubled to 29 tons, valued at 795,852 francs, while in the following year Djibouti imported 25 tons of khat, worth 344,244 francs. Some of this khat was imported from Yemen, and some was smuggled from Ethiopia through British Somaliland by Somali traders. Khat, was imported to British Somaliland illegally across the border because since 1921 khat had been banned in

[45] Richard Pankhurst states that, around the First World War, khat exports from Harer were about 30 tons a year. *Economic History of Ethiopia*, p. 204. To support his assertion, Pankhurst cites numerous sources, although none of them suggests any khat export from Ethiopia, let alone giving actual numbers. Curiously, most of the cited authors visited Ethiopia and had their accounts published long before 1914. To cite a few, Harris, *The Highlands of Aethiopia*, vol. 3 (1844), pp. 423–4; Moktar, 'Notes sur le pays…' (1876), p. 269; L. Robecchi-Bricchetti, *Nell' Harar* (1903), pp. 129–33; J.W. Jennings and C. Addison, *With the Abyssinians in Somaliland* (London, 1905), p. 54.

[46] Adem Aliyu Jaylan, 61, AwWoday, January 19, 2000

[47] René Hachette, 'L'Importance de la Côte Française des Somalis,' *Revue Géographique et Industrielle de France*, 2 (1936), in AOM, Carton 341 Dossier 1.

Table 3.1 *French Somaliland (Djibouti): Khat imports,* 1929–32

	1929		1931		1932	
	(vol.)	(value)	(vol.)	(value)	(vol.)	(value)
Somaliland	6048	109087	479	9867	–	–
Aden	189	3894	18793	392997	1560	16087
Ethiopia	6854	209774	9830	392998	3379	83430
Total	13091	322755	29102	795852	25076	344244

Source: Compiled from *AOM*, Carton 341 Dossier 1, Carton 342 Dossier 7 and Carton 346 Dossier 11. Volume is in kg and value is in French francs.

the British territory. Traders nonetheless continued to choose this route to avoid the high cost of train transportation.[48]

The small quantity of exported khat was carried from the farms on the Harer highlands to the Dire Dawa morning market by individual farmers who traveled the distance on foot overnight with headloads of khat. In the late 1920s the amount of khat export increased following construction of roads that connected several khat production areas to the railway terminal. Beginning in the mid-1920s, *Ras* Tafari took a personal interest in promoting road building in several areas of the empire, but mostly in Harerge.[49] The paving of the important Harer–Dire Dawa road was completed in 1931, while the Ethiopian side of a track connecting Harer to Hargeisa was completed in 1932. By 1934 it was possible to travel by car from Harer to Dadar in the Chercher district. Feeder roads were also completed connecting the Harer–Dire Dawa road to districts such as Gara Mula'ata and the Harer–Asba Taferi road with Bedessa, Hirna and Kunni.[50] The roads were used by expatriate merchants who resided in Dire Dawa and Harer mainly to transport coffee to the railway station by wheeled carts. Yemeni coffee traders reasoned that a lucrative khat business was possible, and they purchased from farmers who brought their product to Dire Dawa and, after processing it, exported the leaves by train to Djibouti and thence to Aden.[51]

Although in the early 1930s khat had become a major cash crop in Harerge, the volume of its export was relatively small. Several reasons contributed to the diminished level of the khat export business. The major roads were built and owned by Emperor Haile Sellassie, a man whom Sir Sidney Barton, British Minister in Addis Ababa in the early 1930s, described as 'better at building roads than at putting them to the best use.'[52] The emperor's representatives demanded exorbitant tolls,

[48] Archives d'Outre Mer (AOM), *Importations, Colonie de la Côte des Somalis,* Carton 342 Dossier 7 and Carton 346 Dossier 11.

[49] FO 371/16996, Barton to Simon, Ethiopia: Annual Report, 1932, January 12, 1933.

[50] FO 371/16101, Barton to Simon, Ethiopia: Annual Report, 1931, January 10, 1932; FO 401/34, Barton to Simon, March 31, 1934; FO 401/24, Dunbar to Chamberlain, March 23, 1928; FO 401/34, Barton to Simon, February 18, 1933.

[51] Interview: Yassin Usman, Dire Dawa, February 20, 1994.

[52] FO 371/16101, Barton to Simon, Ethiopia: Annual Report, 1931, January 1932.

thus causing the two or three minibuses that existed then to charge passengers as much as 10 Ethiopian dollars for a single journey between Harer and Dire Dawa. The high cost made automobile transport unaffordable for Ethiopian traders. The Harer–Jijjiga–Hargeisa road had been built and owned by a certain Lij Mesfin Keleme Work and Co. and only the Indian merchant Mohammadally could afford to have his trucks use the road.[53] Moreover, roads and bridges were frequently washed out during rainy seasons, and the export of fresh khat consequently remained constricted throughout the first three decades of the twentieth century.[54] By the late 1920s and early 1930s, the evidence shows, a new era had arrived for khat even as its importance as an export product grew somewhat unevenly. It is only logical to anticipate changes in the agrarian system and the choices and options available to Harerge's farmers as they devised strategies to cope with changes in and around Harerge's farming communities.

[53] FO 401/30, Barton to Henderson, March 31, 1931.
[54] FO 401/30, Barton to Henderson, December 31, 1930.

Four

Production

The agrarian system, population growth & farming strategies 1935–64

> Coffee is useless. Our fathers cultivated coffee because, in their days, there was no foreign market for khat, and they all died of lice [of destitution]. None of them had a single shirt. We, their sons, grow khat and live like *ferenjis* [Europeans].
>
> – Hajji Abdi Bakar, 1994

In the late 1930s and early 1940s, agriculture in the Harer highlands involved field cropping, horticulture and cultivation of perennials. Cultivation of grains (mainly sorghum), fruits and root and leguminous crops was the economic mainstay of the population. Coffee was the main cash crop and constituted more than half of Ethiopia's total export until the mid-1920s. Harerge's share of coffee exports began to increase in the 1930s and in the subsequent decades, despite rising rail charges, inadequate farm-to-market transportation systems and a lack of direct access to world markets.[1] During the Italian invasion, coffee production fell drastically, only to rise again after the liberation. International market prices and conditions remained favorable throughout the 1940s, thereby helping coffee to dominate the export-commodity sector of Harerge's economy.

In the early 1950s, coffee prices fell and production started to decline. Farmers began to replace their coffee trees with khat.[2] The latter quickly gained a sizeable market in Harerge's urban centers as well as in Aden Colony, and French and British Somalilands. By the early 1960s, khat not only had become a major part of the cropping system in Harerge, but also accounted for 5.3 percent of Ethiopia's total export returns, ranking fifth among all commodity groups. It furnished a large percentage of government revenue and contributed almost 50 percent of the total cash income of farming families. The khat business provided

[1] Santagata, *L'Harar*, pp. 33–78.
[2] Darby *et al.*, 'Khat,' p. 169.

employment for a considerable number of the inhabitants of the major urban centers. By the early 1960s, khat was an important cash crop on which farmers, traders and the government depended for income.

This chapter seeks to analyze khat's transformation from a limited domestic crop to a cash crop sold in local markets and, then, to the most profitable export commodity. Our questions here are those the farmers, traders and consumers asked themselves. How can I make the best use of the land that I have? What cultivation pattern and cropping balance would enable me to provide for my family? How can I increase the yield from my land to keep abreast with the growing number of my family? These seem to be rather simple queries, matters for straightforward answers, but therein lies the main problem of this study.

The difficulty is that the material needed to answer many of these questions is inadequate. Interview materials are fragmentary, nonchronological and often silent on important issues. Quantitative data are not available, and the few statistics that exist are notoriously inaccurate. Administrative records lack information relating to agriculture, land-use patterns and population distribution. Aware of the limitations, I will attempt to extrapolate answers from various data to the central question of khat production in Harerge and argue that population growth, the expansion of rapid transport and communication facilities and the monetization of the regional economy were responsible for the commodification of khat. This argument shifts emphasis from the notion that market forces alone spurred the commodification of khat and proposes a more balanced approach that looks at change from the perspectives of production, consumption and marketing. The analysis will begin by describing the agrarian system and by delimiting the place of khat in that system in the late 1930s and early 1940s, and then moves on to an explanation of how the limitations on the reproductive capacity of the agrarian system led to the expansion of khat cultivation, paying special attention to farmers' responses to land scarcity and the advantages of cultivating khat.

The changing role of khat in the agrarian system

When analyzing the process of khat's commodification and the place the crop has acquired within Harerge's agrarian system, it is essential to bear in mind two fundamental issues regarding the plant's ecology. First, the khat plant thrives best in well-drained soils at 5,000 to 8,000 feet. Situated within this elevation range, the Chercher (western) highlands, the Gara Mula'ata massif and the Harer (eastern) highlands have always been major centers of khat production. In these regions, producers consumed nearly half of the khat they harvested and sold the rest to urban dwellers. Except for the Hareri, most of the consumers were ethnically and linguistically distinct from the producers and lived chiefly in the semi-arid Shebbelle and Awash plains. Thus, khat historically has been exchanged across ecological frontiers that have also been cultural frontiers.

Second, khat is well suited to areas where perennials and annuals also grow best. One of the major perennials, coffee, unlike khat, is tied to world market structures and has been an important source of foreign exchange. As a result, successive Ethiopian governments have preferred to encourage the production of coffee as the main cash crop and food crops for consumption over khat. From the standpoint of local farmers, however, khat is not a peculiar leaf used by a few 'addicts,' but an important crop in their farming system and a commodity grown for consumption and the market. Over the last four decades, economies of scale and returns on investments have favored khat, and Harerge's highland farmers have increasingly turned to its production on a large scale.

In the late 1930s and early 1940s, Harerge's agricultural economy centered on the cultivation of sorghum, which occupied greater acreage than other cultivated plants and yielded more than the combined total of all other grains.[3] The crop owed its popularity to its adaptability to a variety of ecological conditions, the relatively simple cropping techniques necessary for its cultivation and its multiple utility. The grain is used as the basic food, the stalks for animal feed and for building small structures, and the roots as fuel, while the remaining parts of the stalk and leaves were scattered on plowed fields as fertilizer.[4] Relatively little of the grain was sold, since Oromo farmers preferred to store any surplus as insurance in case of crop failure. Despite price variations and fluctuations, a quintal of sorghum during the harvest in November and December sold for about 7.50 birr (at the time US $1 = 2.50 birr) at the threshing ground. A month later the price sometimes climbed to 10.00 birr, and in June and July a quintal sold for 15.00 birr, with better prices in Dire Dawa if farmers had donkey transport. Corn, the second most important crop, was normally sown with sorghum in April and May, but one fast-growing cultivar was planted in early June in gardens or in late May with fast-maturing varieties of sorghum. In most farms, corn occupied less of the cultivated fields and the average yield was low, probably 2 to 3 bushels per acre. In a good year, sole-cropped corn yielded 15 to 25 bushels per acre. Because corn was more productive in higher and wetter areas than in lower and drier regions, yield as well as the acreage rapidly decreased above 7,000 feet. Like sorghum, corn was used for human consumption and its stalks for animal feed, construction, fuel and fertilizer.

[3] In 1935, traveling along the Harerge–Dire Dawa highway, Colonel Comyn-Platt observed that 'there were miles of cultivation together with flocks and herds that occupy a scattered population to be seen on either side. They are the Gallas (Oromo), the best agriculturalists in the land [who] have lived and toiled in these parts, the cradle of their birth.' T. Comyn-Platt, *The Abyssinian Storm* (London, 1935), p. 71; Santagata, *L'Harar*, pp. 33–61.

[4] Leonard Miller and Telahun Makonnen, *Organization and Operation of Three Ethiopian Case Farms,* Imperial Ethiopian College of Agriculture, Experiment Station Bulletin No. 35 (Dire Dawa, 1965), p. 19.

The association of these two crops allowed for distribution of risks and extended availability of food and fodder. Sorghum suffered from none of the diseases that afflicted other cultivated grasses, but was often damaged by severe frost. Corn was harvested in October, before the frost season sets in, thus sparing a farming household from losing the entire annual harvest. When, on the other hand, a drought or exceedingly heavy rains occurred, corn was damaged but sorghum survived. Furthermore, corn provided the farming household with food at the end of the rainy season, when stored sorghum had been used up. More preferred as food, the harvested sorghum was stored for the remaining part of the year. To feed livestock, over-sown corn and sorghum fields were regularly thinned, providing daily green fodder during the rainy season. Once dried, corn leaves rotted and disintegrated rather quickly, thus sorghum leaves were used as fodder until the next season.[5]

Barley and wheat were important highland crops at elevations above approximately 7,500 feet. At lower areas, they occupied relatively small acreage, chiefly small fields and plots, planted late in the sowing season after a partial failure of sorghum. As rotation crops, barley and wheat were usually planted in June and harvested four months later in October, a full month or two before the major sorghum harvest. In lower elevation areas, they served as substitutes when the two coarse grains failed to yield owing to soil exhaustion, catastrophic damage to young plants or pests. Moreover, when fields were left unsown in April and May, owing to labor scarcity, farmers put in wheat and barley in July and collected the crop before the sorghum harvesting season.

The major leguminous crops common to all villages in the Harerge highlands were fenugreek and beans. Their importance lay in their early maturity and low moisture requirements. These legumes were interplanted with sorghum and corn and required no additional labor investment in seed bed preparation, hoeing and thinning. In most areas, two harvests were taken between April and December. Most of the fenugreek was used in food preparation, and beans were sold for as much as 15 birr per quintal. Very little fenugreek was sold. In some areas, especially to the east and south of Harer, peanuts were planted with grain sorghum and corn, and most were sold in markets in small and major towns.

Farmers planted perennials such as coffee and khat. In the 1940s rising demand made coffee a cash crop of choice, and the good price of those years enriched producers. In the immediate decade before the Italian occupation, Ethiopia's annual coffee export averaged 15,000 tons and showed a steady rise in volume, reaching 18,000, 19,000 and 20,000 tons in 1934, 1935 and 1936, respectively. The increase in value considerably exceeded the increase in volume figures obtained in those years,[6] and

[5] Clarke Brooke, 'The Durra Complex in the Central Highlands of Ethiopia,' *Economic Botany*, 12:2 (1978), p. 199. The research for this article was done in the early 1950s as part of the author's dissertation fieldwork.

[6] FO 371/53561, Cook to Bevin, May 22, 1946.

rising prices led to higher production in Harerge, particularly in the Chercher highlands.

The occupation years saw a precipitous decline in coffee production. In fact, there was no export at all in 1938 and 1940.[7] Following the liberation in 1941, coffee exports began to rise in response to high wartime prices and demand. Harer's coffee in particular benefitted from the Middle East's willingness to pay elevated prices. Consequently, Ethiopian coffee export reached a total of 2 to 3 million sterling in 1944–5, and much of this came from Harerge. High wartime prices provided farmers the incentive to increase coffee production and made coffee business so attractive that high-ranking Ethiopian government officials, including Emperor Haile Sellassie, started to engage in it.[8]

Throughout the 1940s, coffee remained first in importance among the cash crops of the Harerge highlands. But not all regions of Harerge benefitted from the coffee boom. In the pre-war period, coffee had enjoyed a similar status in the densely populated Harer plateau as in Chercher. During the early days of the occupation, however, the Italians introduced price controls, which discouraged the small cultivator from increasing or even maintaining previous levels of production. In 1938, Emilio Scarin reported seeing extensive fields of unharvested, ripened coffee in the vicinity of Harer where thousands of acres were cleared of coffee trees and planted with cereal and other crops. Much of Harer's coffee, which constituted a significant portion of Ethiopia's coffee exports, came from the Chercher highlands.[9] One legitimate question in this regard is why farmers in the Harer highlands did not revert to coffee production to participate in the expanding market of the 1940s. The existence of coffee trees that are about forty years old around Harer suggests that some farmers did start planting coffee in the mid-1940s. Nevertheless, since it took at least five years for coffee trees to yield harvests, the time span of less than a decade was not sufficient for the Harer highland farmers to have joined the coffee boom, for the price slump started in 1951.

Before the Italian invasion, khat was a part of the cropping system of the Harerge highlands, but it did not occupy a significant portion of the crop field. In Chercher, khat was often bench-terraced and always planted on the contours, especially in the area between Dadar and Hirna.[10] The Christian 'Amhara' settlers, as noted earlier, despised khat-chewing and entered into sharecropping agreements only with those farmers who rented their lands to cultivate grains. Muslim Oromo farmers grew their own khat and sold small amounts in Harer town and

[7] R. Pankhurst, 'A Chapter in Ethiopia's Commercial History: Developments during the Fascist Occupation,' *Ethiopia Observer* 14:1 (1971), p. 64.

[8] FO 371/80232, Reports from Harar Consulate by Major Pink, March 9, 1950. A German national, Frederick Hall, who was employed by the emperor, installed a coffee-cleaning machine in Harer and the factory functioned under monopoly rights to process all the coffee produced in the province.

[9] Scarin, *Hararino*, p102.

[10] FO 371/53461, Scrivener to De La Warr, March 15, 1946.

in Dire Dawa, some of which was exported to Djibouti and Aden.[11] Domestic and foreign demand apparently remained low throughout the early 1940s, owing to wartime disruptions. Farmers had very little incentive to expand their khat plantations. Whatever demand existed in urban centers was satisfied by production in the Harer highlands, not in the more distant Chercher. By the early 1950s, as we shall see in the next chapter, improvements in transportation facilities in the region had made khat a very competitive and viable cash crop, which farmers preferred to coffee.

Unlike in Chercher, where coffee was planted extensively, perennials did not occupy large portions of acreage and labor in the Harer highlands. Most farming families were inclined toward producing food crops for household consumption. They planted mostly tree crops to generate cash to cover taxes, clothing and other nonfarm goods. This strategy worked well as long as there was enough land to which cultivation might be extended. By the 1940s, however, arable land was not sufficiently available. The land tenure system that had evolved over more than five decades had reduced the size of a family plot to the extent that expanding the cultivation of both annuals and perennials was nearly impossible. The limitations on maintaining the existing cropping patterns obviously precipitated a crisis within the agrarian system. The expansion of khat cultivation in the 1940s was the result of farmers' response to the problem of land scarcity.

Population growth & land scarcity

In 1941 Harerge had a population of approximately one million.[12] The distribution of population was conspicuously uneven, owing largely to the geographical and topographical characteristics of the province. The Shebbelle and Awash plains were sparsely populated by various pastoralist Somalis who occupied vast areas with their cattle while the Harerge and, to a lesser extent, the Chercher highlands were densely settled by agriculturalists. Thus, the distribution of population did not relate to the potential productivity of the land. The latter was plentiful where population densities were low, whereas labor abounded where land was scarce. The resource imbalance worsened with natural population increases, presaging adverse effects for the agrarian system.

Census data are not available for a quantitative analysis of the relation between demographic changes and agricultural production in Harerge in the 1940s, but there are various estimates that could help paint a hazy picture of population densities. An Italian study of the

[11] Santagata, *L'Harar,* pp. 67–8.

[12] According to Santagata's estimate of 1938, Harerge's population was 1.5 million. If we subtract the population of Arsi and Bale, which he estimated at 185,000, and included in Harerge, we get a total population of 1.2 million for Harerge. See Santagata, *L'Harar,* p. 92.

Harerge highlands, published in 1942, puts population density in the Harer plateau at 60 persons per square kilometer and in the middle-range altitudes at 100 persons per square kilometer.[13] Italian figures must be treated with caution because most have tended to minimize Ethiopia's population. Compared to later estimates, this particular Italian figure for Harerge seems an obvious underestimation. Calculating on the basis of figures obtained from voter registration for the first parliamentary elections of 1956, an Ethiopian geographer, Mesfin Wolde Mariam, argued that Harerge's rural densities were over 100 persons per square mile in some areas. This estimate, which the author believes is conservative, would mean that population in Harerge had doubled in little more than a decade.[14] However, since Ethiopia's annual rate of population growth in the 1940s and 1950s was only 1 percent, Harerge's population could not have doubled in a matter of a decade.

Mesfin's projections were corroborated by estimates given by Clarke Brooke, an American geographer who conducted field work in Harerge in the early 1950s. His study indicated a relatively dense population on the plateau areas lying between 5,500 and 8,000 feet; Harerge's average rural population density was 150 per square mile, with local variations from 75 to 400 per square mile. Areas estimated to be most densely populated, that is, greater than 150 per square mile, were situated between approximately 5,800 and 7,200 feet, specifically in Karsa, Haramaya, Gara Mula'ata and Kombolcha districts, the vicinity of Fedis, and the areas east of Harer as far as Kocher, west of Jijjiga. The average density of rural population in the Chercher highlands was approximately 75 per square mile.[15]

Although these estimates were not based on accurate census data, they point to the fact that Harerge's rural population had greatly increased in the ninety to a hundred years before the 1940s, corresponding to the transformation of the Harer Oromo from livestock herding to sedentary agriculture. Studies in other parts of Africa indicate that the impact of a growing population density on the pattern of land use becomes apparent when population rises to a point where farming communities enlarge their cultivated areas to the extent that there is no longer sufficient land to permit effective bush fallowing. It has been suggested that this may be expected to occur when the density reaches 150–220 per square kilometer.[16] According to our evidence, over extensive areas of the Harer highlands this density was reached in the late 1940s.[17]

13 Scarin, *Hararino*, p. 95.

14 Mesfin Wolde-Mariam, 'An Estimate of the Population of Ethiopia,' *Ethiopia Observer* (1961), p. 138. Censuses taken later repeatedly confirmed the accuracy of Mesfin's estimates.

15 Brooke, 'The Rural Village,' pp. 61–2.

16 A.T. Grove, 'Population Densities and Agriculture in Northern Nigeria,' in K.M. Barbour and R.M. Prothero (eds), *Essays on African Population* (London, 1961), pp. 115–36.

17 According to one study, by the 1980s, population density in the Haramaya area had become exceptionally high, 440 people per sq. km and a total of 2,000 families with an

By the 1940s nearly all of the Harerge Oromo had become agriculturalists and occupied vast areas of Harerge. Sedentarization put pressure on the resource base of the region, especially on land. New land for crop production and livestock management, once sufficiently available nearly everywhere in the highlands, was sparse in most parts of the zones favorable for agriculture.[18] There were changes in land-use management, relative variations in the size of holdings and even absolute land shortages in some districts. These changes were symptoms of high population density, as the result of which land became more limited in relation to labor.

Aggravating the problem of diminishing family plot size were the landholding and inheritance systems. We have seen in Chapter 3 that allocation of land to settlers, the Orthodox church, and civilian as well as military officials had decreased the land available to the Oromo farmers. In the late 1930s, there were an estimated 300,000 new settlers in Harerge to whom land had been assigned. It is not difficult to imagine the severe strain immigration would bring on arable land and on production levels, but additional pressure was caused by the land privatization scheme introduced by the restored Ethiopian government. Intent on increasing revenue and enlarging its political constituencies, the government assigned land in lieu of pensions (*metoriya*) to individuals who had served during the war and in various government capacities. Others, who had been previously granted government land, converted their holdings to freehold, resulting in concentration of land in a few hands.[19]

Between 1947 and 1952 new measurement schemes were introduced and lands that had previously been divided into measured (*qalad* or *gasha*) and unmeasured (*shibetta*) divisions were converted to hectares (1 *qalad* = 40 ha). A detailed analysis of land-ownership statistics in the Harer highlands shows that a total of 0.1 percent of the total hectares of privately held land was shared by 21 percent of the total number of owners, with an average size of 0.39 hectares for each holder. About 3.3 percent of the land was shared by 23.3 percent of the total private holders, with an average size of less than 40 hectares, while roughly 44.6 percent of owners possessed 13.4 percent of the total land. More

[17] (cont.) average family size of six persons. Milkessa Wakjira, 'An Economic Analysis of the Development for the Alemaya Lake Sub-watershed with Special Reference to Soil and Water Conservation Methods,' M.Sc. thesis, Department of Agricultural Economics, Alemaya University of Agriculture (July 1989), p. 39.

[18] Interviews: Mumed Ahmed Umer, Ganda Issa, Haramaya, May 17, 1994; Hajji Mume Kiko, Bate, Haramaya, May 20, 1994.

[19] Under the new system, private lands comprised several types of holdings. The main categories included the *gebar meret*, land over which the holders exercised the right to transfer either through sale or inheritance; *maderiya* lands, originally granted by the government in lieu of salary and recognition of public service but which holders gradually converted to freehold; and the *rist–gult*, lands which were generally granted to dignitaries with distinguished service to the government. Lands which the Ethiopian Orthodox church acquired through grants and transfers were generically referred to as *semon* and were held as private property. Any property that was not classified under one of the two main categories was by definition government property.

interestingly, 25 persons alone, or 0.2 percent of the total number of landowners, possessed 74.6 percent of all cultivable land (measured and unmeasured) or a total of 1,245,915 hectares.[20]

Table 4.1 A *summary of land ownership distribution in Harerge Province,* 1952

Range in size (ha)	No. of owners	Total land owned (ha)	Average holding size (ha)	% of all owners	% of total land owned
< 5	2,633	1,042	0.39	21.0	0.1
5–30	2,915	55,060	18.8	23.3	3.3
40	5,585	223,400	40.0	44.6	13.4
50–120	1,131	89,700	79.3	9.0	5.3
130–360	206	40,510	196.7	1.6	2.4
400–600	32	15,400	481.2	0.3	0.9
> 600	25	1,245,915.2	49,836.8	0.2	74.6
Total	12,527	1,671,027.2	133.3	100.0	100.0

Source: Adapted from Seleshi Wolde-Tsadik, *Land Ownership in Hareerge Province*, 1966.

The concentration of land reflected in the total land owned in Hareerge is also evident in the unmeasured or *shibetta*[21] holdings. Under the new measurement scheme, land that was previously designated as *shibetta* (private holding) was renamed *kuter geber* land. This type of land was scattered throughout the Hareerge highlands, just as were the measured hectare lands, but the *kuter geber* was more typical in the Harer highlands than in Chercher, where the average holding was considerably larger. As in the case of hectare land, *kuter geber* was also concentrated in few hands. An analysis of the *kuter geber* land shows that 95.4 percent of the land owners possessed less than 1 *kuter geber* of land, of which only 25.4 percent had an average holding amounting to 0.5 *kuter geber*. A mere six people, or 0.01 percent of the owners, commanded close to 64.4 percent of the total *kuter geber* land. Of these, one person alone had a share of about 99.3 percent, or 63.9 percent of the total *kuter geber* land, which is held in the form of *rist-gult.*[22]

That only a few individuals owned too much land in Hareerge had two closely related implications for agriculture. Absentee landlords controlled access to a significant part of the arable land, estimated at close to 25 percent of the total arable land at the time. Second, as population

[20] Seleshi, *Land Ownership*, pp. 18–19.

[21] A term introduced by the Egyptians during their occupation to designate plots of land of varying sizes privately held by families. One *shibetta* is equivalent to 4.3 hectares. In the 1960s four to five families shared one *shibetta* or *kuter geber* land, averaging less than 1 hectare per family. Demissie, *Kottu of Ethiopia*, pp. 71–3.

[22] Seleshi, *Land Ownership*, pp. 13–20.

increased, family plot sizes diminished and arable land scarcity became a major factor farmers had to take into account in farming strategies. In most cases, landlords rented out their lands under tenancy agreements, which took several forms of sharecropping. Yet because their land was also a source of fodder, timber and fuel, landlords were not always eager to employ sharecroppers. They derived income from selling resources, often letting local people use their land to graze their animals for charges of 6 to 10 birr per cow. The landlords' preferences for cash income from such practices led to severe shortages in land available for cultivation from which farming families could earn their subsistence.

Just as the concentration of land in a few hands limited access to land, the Oromo inheritance system also adversely affected land use and agricultural productivity. The Oromo were patrilineal, with property passing from fathers to sons (or daughters in extreme cases). Sons took precedence over daughters, though the latter were not entirely excluded from inheritance. When a holding was to be divided, the portion received by daughters was much smaller, often one-sixth that of sons. Normally at death, therefore, land was transferred to the heirs of the cultivator. This was probably a residual effect of a communal system of landholding that had governed the pastoralist Oromo and those who lived beyond the administrative reach of the emirs of Harer.

In a society where population growth rates are high, a partible inheritance system frequently results in a break-up of land since the inherited land must be shared between sons at the farmer's death. A common consequence is yield reduction, which often leads to a subsistence problem for the entire agrarian system. Historically, pre-industrial farming societies have responded to scarcities of arable land and the subsequent reduction in yield either by increasing productivity or limiting the rate of population growth.[23] Harerge's farmers opted for the former.

Farmers' responses to land scarcity

In Harerge, farmers attempted to increase productivity by augmenting the cultivated fields through clearing vegetation or, to put it in the farmers' words, 'claiming land from the forest.' It is probable that woodland was at one time much more extensive in the Harer highlands than it was in the 1940s. The growth of the sedentary population with its increasing needs for crop land, woods for structures and fuel, and the practices of unrestricted browsing reduced the wooded areas to patches on the plateaux between 7,000 and 8,000 feet elevation where most of the population lives.[24]

If we take the district of Kombolcha, for instance, land clearing was a longstanding phenomenon that seems to have intensified in the 1940s. Terracing and manuring were practiced to a limited extent, but, in

[23] David Grigg, *Population Growth and Agrarian Change: An Historical Perspective* (Cambridge, 1980), pp. 29–47.

[24] Brooke, 'Settlements,' pp. 171–2.

Photo 4.1 *Khat terraces, Dhangago hills, Haramaya*

Photo 4.2 *Khat terraces, Faldhaana, Kombolcha*

nearly every settlement, land clearing for cultivation purposes was in far greater evidence than the techniques of preservation. Large parts of the hill areas and the slopes are today covered with very old terraces, which are currently abandoned or left to fallow. A visual comparison of the landscape today with aerial photographs taken in the early 1960s shows that the landscape has not changed significantly over forty years, clearly indicating that most of the land-clearing activities in this area were completed by the early 1960s. The normal cycle of fallowing before land is cleared again for cultivation ranged from fifteen to forty years. This means, land clearing started before the 1940s and significantly declined, if it did not stop altogether, in the early 1960s.[25]

Where vegetation clearing was not possible, such as when landlords limited access to protect specific needs such as timber, farmers increased their own cultivated fields through reduction of the fallow areas. This practice was prevalent in the 1950s and was most intensive in the regions surrounding Harer town. Studies conducted in the mid-1980s on the farming systems of Tula and Kako, two villages north of Harer, show that, thirty years before, fallow areas used to represent approximately 26 percent of the total arable land there. In the early 1980s, fallow land dropped to 10 percent of the arable area in Kako and 22 percent in Tula. Taken together, fallow areas in the two regions fell by an average of 15 percent in forty years.[26]

Studies from other African regions show that, after the disappearance of reserve land, further intensification must take the form of increased inputs of manure or fertilizers, labor or changes in farming practices in order to keep production abreast of population increases. But, for intensification to take place, it has been suggested that population density must reach a critical level of 200–250 persons per square mile.[27] This point was reached in some places in the Harerge highlands in the 1950s, but intensive cultivation was not much in evidence there, probably owing to the lack of increased inputs. Manure resources fell below the level necessary for maintaining soil fertility because of the decline in the number of livestock caused by reduced grass fallowing. The end result was a significant drop in production. Harerge's farmers, showing little tendency for outmigration, resorted to deforestation whenever such was possible, parceling of farm lands and farming on marginal lands that owners did not need for non-farm resources. Their response to diminishing plot size, land scarcity and declining production was extensification of cultivation in areas where land was available.[28]

[25] Georges Risoud, *Evolution of Peasant Agriculture in the Eastern Harerge Highlands: Development Prospects,*' Farming Systems Research, Alemaya University of Agriculture (Alemaya, 1987), p. 17.

[26] Ibid., p. 22.

[27] W. Allan, *The African Husbandman* (London, 1965). See his formulation of critical population density.

[28] Demissie Gebre Michael, *Land Tenure in Bate*, Imperial Ethiopian College of Agriculture, Experiment Station Bulletin 49 (Dire Dawa, 1966), pp. 7–14. Central Statistical Office, *Statistical Abstract* (1972), p. 32.

Intensive cultivation was limited to dry-season production. Beginning in the early 1960s, farmers turned to cash crops owing to their ability to irrigate their plots during dry seasons. Dry-season production was for the most part limited to the production of perennials. In the 1960s, it was not unusual to see irrigated vegetable gardens, fruit and coffee orchards in the vicinity of Harer, but generally irrigation was used in khat production, which, as we will see, was more profitable by any measure. Irrigation was made possible by water pumps, which some farmers were able to purchase from khat profits and rented out to other farmers whose farmsteads were proximal to a lake or a big stream.[29]

The role of irrigation in cash-crop production at this time need not be exaggerated. Farmers usually grew perennials only on the land they owned and when such lands were close to water sources. On rented lands, tenants are said to have had more say in crop choices as long as the crops to be raised were annuals. In most cases, the preferences of both the landlord and the tenant were the same, and decisions were made by the tenant with little interference from the landlord. However, the consent of the landlord was critical if the tenant were to decide to plant perennial crops such as khat. Such tenancy arrangements were rare, if at all they existed, for landlord–tenant agreements would be complicated where, at the end of the lease period or for other reasons, the tenant asked to leave the land without compensation.[30]

The three strategies outlined above were essentially temporary tactical responses to a long-term problem: extensification was an effective solution only as long as sufficient arable land existed; intensification required inputs such as fertilizers that were not readily available; and irrigation solved farming problems only where there was enough water. To put it simply, these strategies eased or put off the problem of land scarcity but did not eliminate it. The agrarian system remained under severe demographic pressure, despite the overall augmentation of the land area, reduction of fallowing time and some improvements in farming techniques.

The effective strategy proved to be the shift to high cash-income perennial crops. Coffee never recovered its Second World War status as the premier cash crop in the Harerge highlands, even under the favorable market conditions of the 1940s. For the areas close to Dire Dawa and the farming regions within 15 miles of the Harer–Dire Dawa highway, khat became the cash crop of choice for most farmers. Increasing demand for khat from the Horn of Africa and the Arabian Peninsula, which will be discussed in the next chapter, had stimulated production, but the shift to khat cultivation was one in a series of

[29] M.E. Quenemoen, Darrell Watts and Yack Mosley, *The Economics of Irrigation with Small Pumps in the Chercher Highland*, Imperial Ethiopian College of Agriculture, Experiment Station Bulletin 58 (Dire Dawa, 1968), pp. 25–43.

[30] Lemma Arity, 'A Regional Study of Hara Maya Based on the Studies of the Alemaya College of Agriculture,' B.A. essay, Department of Geography, Haile Sellassie I University (1969), p. 47.

decisions that farmers made at the farming-system level in response to the limitations population pressure imposed on the agrarian system. To substantiate this assertion we may need to look at two areas where the agrarian landscape has changed completely as the result of expanded khat cultivation. These are Tula and Kako, located about 5 miles to the north and 10 miles to the northeast of Harer, respectively (Map 4.1).

In Tula, khat had been under cultivation for a long time. Nineteenth- and twentieth-centuries travelers consistently reported observing khat fields in the district. In the early 1940s, as farmers did elsewhere in Harerge, Tula farming communities devoted much of their land to the cultivation of sorghum, corn and other grains but grew some khat for consumption and for retail in Harer market. Interviews with elderly farmers indicate that, in the environs of Harer, the long-term process by which the customary sorghum–maize cropping pattern was gradually replaced by khat plantations set during the early 1940s.[31]

One study shows that the cultivated land in Tula belonging to six farming families increased by 36 percent between the time of inheritance (1940s) and the 1980s. In 1986, khat covered 40 percent of the arable land and 68 percent of the total cultivated area. The shrub was sole-cropped on approximately 10 percent of the total cultivated area, whereas it grew on 50 to 55 percent of the reclaimed hillside areas. If one considers the total fields under cultivation, hillsides or other areas, khat was grown in association with other crops on more than 80 percent of the planted areas.[32] A similar study of the cropping patterns at Tula shows that farmers began to plant khat trees on land that used to be under food crops in the 1940s. The substitution process lasted less than two decades. From the mid-1950s onwards, Tula farmers continued to expand khat plantation but at a much slower pace and in association with cereal and root crops.[33]

In Kako, the story is slightly different. According to oral information, khat was not cultivated there on a significant scale until the 1920s. Coffee was the main cash crop, reaching its production peak in the late 1930s. A few farmers still own some coffee trees planted by their parents, but in general most have been replaced by khat. A study of the cropping patterns of eleven farming households in Kako shows that the area covered by khat increased by an average of 55 percent in a matter of one generation. At the time of inheritance or the time the current generation established families (the 1940s), khat constituted 17 percent of

[31] This replacement scheme was carried out by the generation that preceded my informants who established a family approximately fifty years ago. The present informants inherited land with some khat at the time of marriage but, rather than maintaining subsistence farming, increasingly shifted to cultivating the more profitable khat. Interviews: Mumed Ahmed Umer, Ganda Issa, Haramaya, May 17, 1994; Mussa Botto, Ganda Gole, Haramaya, May 23, 1994.

[32] Herve Wibeaux, *Agriculture in the Highlands of Hararghe, Kombolcha Area: Study of Six Farms*, Farming Systems Research, Alemaya University of Agriculture (Alemaya, 1986), pp. 42–4.

[33] Risoud, *Evolution of Peasant Agriculture*, p. 38.

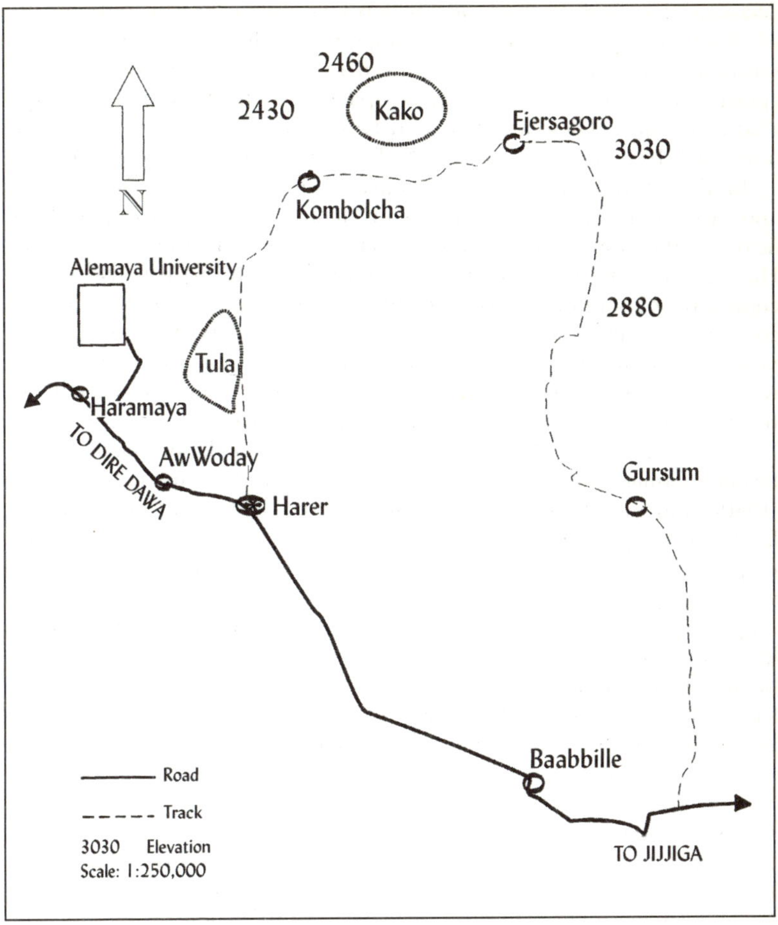

Map 4.1 *Location of Tula & Kako regions*

the arable land belonging to the eleven families. In 1986 khat covered 23 percent of the area under cultivation, indicating that khat plantation grew by 6 percent in less than three decades. The expansion of khat cultivation stopped in Kako from the late 1970s onwards owing to the adoption of potato and vegetable crops.[34]

The difference in the pace of change between Tula and Kako is a function of their geographical location. Tula is closer to Harer, and farmers were able to make daily trips to the main market there. After the completion of the Harer–Dire Dawa road, Tula farmers were able to sell their khat along the highway (see Chapter 5 for the purchasing methods of khat in the 1950s). The selling price was much higher at the roadside markets than in the small local markets such as Kombolcha because the purchasers were agents of large exporters based in Dire Dawa. The effect of Tula's proximity to a highway was in evidence even in the early 1990s in its relative wealth and the prevalence of khat in the cropping system.[35]

The process of change in these two areas brings to attention two important facts about the agrarian system and points to an underlying logic that informs farmers' decisions regarding farming strategies. First, the evolution of cropping patterns, as exemplified in the development of intercropping and the increasing emphasis on cash crops, reveals that most farming families in Harerge had a shared notion of maintaining food self-sufficiency while enhancing the farming system's capacity to produce for market and to maximize cash income. Second, it points to the important role of speedier transportation in the development of khat plantations. It is no coincidence that areas emphasizing the cultivation of khat are those regions that lie along the Harer–Dire Dawa road.[36]

In a broader context as well, the agrarian system shows a general trend toward a gradual expansion of the land committed to khat cultivation. Between 1954 and 1961, the acreage devoted to khat production in Harerge more than doubled, from 7,400 to 17,300 acres; and these figures presumably exclude those farms where khat was intercropped with other staples.[37] As demand slowly grew for two decades following the Italians' expulsion in 1941, farmers extended their khat fields by planting the shrub on hillsides and on marginal lands where erosion was severest. On good arable lands, where erosion was not a major problem, food crops were cultivated by most farming families in order to meet their household needs. Photographs taken in the 1950s show that most hillsides in the districts of Haramaya, Kombolcha, Harer, the eastern extremity of the Chercher highlands and the entire area of the Dhangago hills were covered by khat trees planted on terraced benches, while the lower plains were devoted to sorghum, corn, vegetables and grazing fields. Such cropping patterns are evidence that, in the late 1940s and

[34] Ibid., p. 40.

[35] Wibeaux, *Agriculture of Hararghe*, p. 120.

[36] These are districts of Kombolcha, Haramaya and Hundene as well the Dhangago Hills.

[37] Amare and Krikorian, 'Coffee's Rival,' pp. 357, 370.

1950s, the cultivation of khat was extended to marginal lands, not at the expense of food crops.

It is not incorrect to assume that khat was a crop that provided the technical answer to the problem of severe land shortages that had dogged the agrarian production system for a long time. Derek W.A. Peters, a British Colonial Service scientist who visited khat plantations in the Harerge region in the early 1950s, explained why farmers preferred khat as their main cash crop. The first reason was the shrub's proven ability to provide long-term yields. The scientist observed many khat fields where the main crops were obtained from trees that were about thirty years old. Farmers also extended the productive life of khat trees to more than seventy-five years by yearly pruning and protecting them from animal browsing.[38]

The second reason for khat's success was related to the agrarian system's labor allocation patterns. Basically, khat needed little maintenance, taking little labor away from other crops during the annual crop calendar. Normally the cultivation, shoring up and manuring of khat trees occurred during months when the farmers would be idle. The heaviest amount of labor in khat's annual maturation cycle was, as today, guarding against theft at night, a time the farmers do not need for other productive purposes. Only the cultivation of irrigated khat slightly overlapped with seed bed preparation, sowing and cultivation of other crops, although the practice was not widespread in the early 1950s. Where khat was irrigated, the labor loss to other crops was more than compensated for by the extremely high price paid for the leaves during the dry season. As a commodity well integrated into the production cycle of other crops, khat provided farming families with necessary cash income without requiring much additional labor input and without disrupting the agricultural calendar.[39]

Thirdly, khat exhibited characteristics innate to a classic cash crop. It was easily sold and often fetched high profits throughout the year. In this regard, it contrasts sharply with coffee. In the first place, coffee prices were declining in the 1950s, and the crop became less marketable than khat. Khat held a competitive edge over coffee in providing at least two crops a year or up to five crops if irrigation was used. In the case of

[38] D. Peters, 'Khat: Its History, Botany, Chemistry and Toxicology,' *Pharmaceutical Journal*, 169 (1952), p. 16. Similar claims are echoed by the present generation of farmers, who describe khat as an insurance crop. They express gratitude to their fathers, who planted and handed down khat fields to them, and vow to follow in the footsteps of their parents and hand down khat fields to their children. Khat thus allowed farmers to raise cash income over several generations without endangering the subsistence requirement of the family, an almost universal claim of my informants.

[39] For annual cropping calendar and labor usage pattern, see C. Langlais, M. Weill and H. Wibeaux, *Farming Systems Research: Preliminary Report and Fture Programme*, Alemaya University of Agriculture (Alemaya, 1984), appendix. Additional information obtained from informants: Mumed Ahmed Umer, May 17, 1994, Ganda Issa, Haramaya; Abdulahi Ahmed, Ganda Issa, Haramaya, May 17, 1994; Mumed Munan, Ganda Lucha, Haramaya, May 24, 1994.

coffee, a good harvest year was followed by a year of low yield. Put another way, an investment in coffee did not promise an annual return, leaving the farmer unable to purchase non-farm commodities.[40]

Research has constantly shown that farmers believe that khat brings better returns on investment than other crops. Two studies carried out in the early and late 1960s have shown that khat accounted for 30–50 percent of the total annual family cash income and 40–60 percent of the total value of a family's home-produced foods. The study of the latter period was done in a district where khat trees occupied only 13 percent of the cultivated land. When asked what revenue they received from one khat tree, farmers told researchers that they earned at least 10.00 birr yearly, as compared to birr 1.00 from coffee.[41] My informant Hajji Abdi Bakar responded to a similar question in the following way:[42]

> Q: How much coffee and how much khat can you collect from one tree of each?
> Hajji Abdi: In the best year you could harvest 5 kg of coffee and 1 kg of khat.
> Q: And for how much can you sell 1 kg of each?
> Hajji Abdi: When prices are good, we sell 1 kg of coffee for birr 10 and 1 kg of *urata* khat for birr 500.
> Q: So it is more profitable to plant khat?
> Hajji Abdi: Oh! you are comparing a slave to a free man, to tell you the truth. Coffee is slavery, khat is freedom. Look, if we harvest coffee this time of the year, we can come back to it only next year or two years from now. What is the coffee farmer to do in the interim, eat dirt? The khat farmer can harvest every month. Coffee is useless. Our fathers cultivated coffee because, in their days, there was no foreign market for khat, and they all died of lice [they were destitute]. None of them had a single shirt. We, their sons, grow khat and live like *ferenjis* [Europeans].

The fourth advantage of cultivating khat was the shrub's adaptability to a wide range of soil conditions and rainfall patterns. Peters found that khat was cultivated on terraced hillsides, interspersed with coffee trees. On lands that sloped rather gradually, khat was intercropped with sorghum, meaning that, in the late 1940s and early 1950s, khat did not compete for land with food crops, because it was cultivated on marginal lands where major grains did not do well. Intercropping, therefore, allowed for more productive land use in a way that is consistent with the overall farming strategy of maximizing cash income while maintaining food self-sufficiency.[43]

Finally, khat had a remarkable capacity to resist diseases and drought. Peters observed that the trees relied entirely on rainfall, since the use of

[40] Interview: Abdulhafez Yusuf, Ganda Lucha, Haramaya, May 5, 1994.
[41] Amare and Krikorian, 'Coffee's Rival,' p . 357.
[42] Interview: Hajji Abdi Bakar, Ganda Lucha, Haramaya, April 6, 1994.
[43] Interviews: Ahmed Amit, Ganda Gole, Haramaya, May 25, 1994; Khalifa Hajji Ali Boru, Ganda Lucha, Haramaya, May 6, 1994.

irrigation was quite minimal. For khat, the incidence of rain in one season rather than the amount of the annual rainfall was the important element for success. In times of protracted droughts, farmers usually removed all leaves on the khat plant as a measure against excessive transpiration and induced dormancy until the rains came back. Moreover, Peters observed that no disease appeared to afflict the khat tree, even when the adjacent coffee trees were seen dying.[44]

Conclusions

Under most circumstances, an expansion of cash-crop cultivation is explained in terms of demand and marketability. The ability to reach distant consumers might well have been key to the rapid expansion of khat cultivation and marketing, but focusing on market-based explanations alone may be ignoring the rationale of production that motivates farmers at the farm level. Agricultural change is not necessarily a function of forces independent of the farm field. Farmers' decision to shift to a new cash crop must be seen within the context of the advantages and opportunities the new crop offers to improve the agrarian system's reproductive capacity.

Yet this assertion should not be misconstrued to mean that external forces did not play any role. No one will produce a commodity for which there is no demand. Farmers make the choice of what kind of crop they should plant at any particular time based on the realities of the farming system. Demand is not the only factor, but one among many, which influences the decision-making process. Therefore, the argument about the expansion of production must be presented in conjunction with an argument about change in the scope and pattern of consumption that led to more demand that stimulated production. The expansion of the culture of khat consumption is closely related to improvements in the marketing and trading of the chewable leaves. Better transportation facilities came with the railroad that connected Dire Dawa with Djibouti and with the introduction of trucking in the 1930s, especially during the Italian invasion of Ethiopia.

[44] Peters, 'Khat,' p. 16. The disease-resistant nature of khat became even more important after a coffee berry disease (CBD) swept through much of East Africa in the 1970s. Whereas most coffee trees died, nothing happened to khat. Interviews: Dejene Ayele, Coffee Processing Expert, Harer, March 4, 1994; Te'ame Tiku, Coffee Protection Expert, Harer, March 4, 1994; Debebe Assefa, Agricultural Extension Expert, Harer, March 4, 1994.

Five

Consumption & Marketing

Pastime, protest & political economy

1935–64

> It [khat] is held by the Ulema here [Harer] as in Arabia, 'Akl al-Salikin' or Food for the Pious, and literati remark that it has the singular properties of enlivening the imagination, clearing the ideas, cheering the heart, diminishing sleep, and taking the place of food.
>
> – Richard Burton, 1853

Before khat became a commercial item in Hararge in the early decades of the twentieth century, the majority of its users had been Muslims who for religious reasons eschewed alcohol, and chewed khat to stay alert during long nights of prayer. The leaves were safely transported by donkeys and human porters to a distance of a little more or less than fifty miles from the production areas, which effectively became the radius within which they were consumed. In the absence of a rapid means of delivery, the stimulant was used and appreciated by well-off consumers and those who produced it for their consumption until modern transportation made it easier to move khat over greater distances.

In the realm of marketing, the physical properties of the commodity carried more weight in determining value than the state of demand and supply or issues of cultural change that were germane to the discussion of the production and consumption aspects. Khat has a very short 'shelf life.' In this regard, it is unlike other psychoactive substances. For example, kola nuts were consumed in substantial quantities throughout West Africa, especially in Muslim areas, before the advent of modern transport. Carefully packaged nuts were transported a thousand kilometers by animal caravans for a duration of up to three months.[1] Unlike kola, which retained its properties, the active ingredients in khat that stimulated chewers seem to lose their desirable properties when transported for more than two days.

[1] Paul Lovejoy, 'Kola Nuts: The "Coffee" of the Central Sudan,' in Jordan Goodman, Paul E. Lovejoy and Andrew Sherratt (eds), *Consuming Habits: Drugs in History and Anthropology* (London, 1995), p. 105.

Consumption began to increase owing in part to improved transportation systems that had been gradually expanding since the turn of the century. Rapid transport facilitated the delivery of fresh khat and consequently increased demand in the Horn of Africa. This chapter attempts to account for the emergence of khat as the most desired stimulant in that region by analyzing socioeconomic changes in northern Somalia, Djibouti and eastern Ethiopia since the 1930s. It will focus on the construction of road networks and the beginning of commercial trucking, which facilitated the delivery of fresh khat to areas where it had not been available before. Then it outlines the spread of khat consumption within and outside Ethiopia, shifts to discussing the economic implications of and government reactions to a fast-growing consumption and concludes by detailing the struggle to control the lucrative khat industry.

Developments in transport infrastructure

Until the Italian occupation, Harerge's agricultural highlands had not been effectively connected with the main transit town of the region, Dire Dawa. The Harer–Dire Dawa road was initially paved in the 1890s for wheeled cart operation but, gullied by floods, it remained virtually impassable and unused by motor traffic until the 1940s; no commercial vehicles operated between Harer and Dire Dawa. The unimpressive performance of the Dire Dawa–Djibouti railway was arguably due to the lack of a modern transportation infrastructure connecting the production areas of Harerge with the railroad terminus and the major market center.

The lack of transport was particularly detrimental to the khat trade. Djibouti was privileged to receive train-delivered khat, although transporting the leaves to Dire Dawa was still a problem. The main towns in the Ogaden lowlands did not receive fresh khat until they were connected, in the 1930s, by motorable tracks to the production areas. Consumers in Jijiga, which was not close to production areas, received some fresh khat from such areas as Gursum and the environs of Harer town. One territory that later became a major khat consumption center, British Somaliland, also remained out of reach except for a few border towns such as Togochale, Tafari Ber and Awkeref.[2]

This situation began to change during the Italian invasion. Once in control, the fascists wasted no time in launching their program to make the new colony economically productive and attractive for settlement. The most spectacular achievement was in road construction projects. In 1935 Ethiopia's road system consisted of just 35 miles of all-weather road. In only five years, the Italians built 4,347 miles of roads, of which 2,145 miles were hard-surfaced. Harerge was given special attention in the Italian public works program. Improved earth roads were built from

[2] Interviews: Hajji Mohammed Abdurahman, Dire Dawa, March 26, 1994; Hajji Dawd Abdo, Dire Dawa, March 26, 1994.

Addis Ababa through Awash and Dire Dawa to Djibouti, and from Dire Dawa through Harer to Jijjiga and to Hargeisa in British Somaliland. Secondary roads were extended to small towns of the province with a network of roads that connected Asba Tafari, Kombolcha and Grawa to the main highway. An all-weather road connecting with the Harer–Dire Dawa highway was constructed through the highlands, affording Karsa, Kulubi and Dadar access to the major communication routes.[3]

These roads linked various khat-producing areas with the major distribution centers of Harer, Dire Dawa and Jijjiga where farmers sold the stimulant to traders who distributed it throughout the province and to neighboring countries. The linking of Hargeisa with Jijjiga and the Harerge highlands had significant implications for the consumption of khat, which now reached Djibouti within forty-eight hours of harvest with its freshness and potency intact.

Consumption in the Horn of Africa & the Arabian Peninsula

The culture of chewing khat, as we have seen, spread among the highland Oromo farmers only in the early decades of the twentieth century. Whereas we find that khat was used in rural areas to generate energy to do farm labor, in urban centers it remained mainly a leisure habit. In Harer town, khat-chewing had been common, at least since the early twentieth century, and indeed came to be considered a mark of being *ya harer lij* (one who's from Harer). In other highland towns, khat consumption grew with the development of essentially 'peasant towns.' Khat farmers migrated to small villages located along the Harer–Dire Dawa highway to start up new businesses. In the process, the villages grew to towns and served as market centers. In the 1950s, as we will see later, quite a few farmers left for urban centers in order to make room for landless relatives. Former farmers in Haramaya and AwWoday considered chewing khat an indispensable ritual.

Khat and its consumption began to reach the lower regions of Harerge, the Ogaden and the Haud, in the early 1940s. In 1945, according to the British administrator of the then Reserved Areas, khat consumption was reported to be on the increase in the territory he administered. The main consumers were Gogtis, Somalis and Arabs. Both men and women were said to be chewing khat most afternoons.[4]

[3] J. Barker, 'Developments in Ethiopia's Road System,' *Geography*, 59:2 (1974), p. 150; Perham, *The Government of Ethiopia*, p. 182.

[4] CO 323/1885/7, G.H.W. Kitson, Senior Civil Affairs Officer, Reserved Areas-Ethiopia, dispatch # 1238, undated (file year is 1946); CO 323/1885/7, Chief Administrator, Somalia Civil Affairs Branch, to East Africa Command, January 31, 1946; CO 323/1885/7, E.C.S. Reid, War Office, to R.C. Cade, Colonial Office, March 21, 1946.

Spurring consumption in Harerge lowlands was the establishment of modern institutions with salaried civil servants after 1941. Ever since the annexation of Harerge to the Ethiopian empire, the government structure had been staffed by Christian 'Amhara' and remained so for the duration of the imperial regime. These officials were initially disdainful of khat-chewing as a form of entertainment, preferring to spend their leisure time drinking local ale (*tella*) and honey wine (*tej*).[5] Alcoholic beverages were, however, not suitable entertainment in the hot towns of the lowlands and were hard to get in the Ogaden, where the only available stimulant was khat. The local residents spent the hot afternoons chewing khat and socializing, and the civil servants followed suit. Thus, khat-chewing spread to the Christian 'Amhara' population in Harerge as a pastime habit, quite distinct from its long-time cultural and religious significance for Muslims.[6]

Not far from northeast Ogaden, khat consumption in British Somaliland was alleged to be insignificant before the 1930s. In Hargeisa, the main town of the Protectorate, there was reportedly only one person who was known to chew khat.[7] Even though the person is identified by name, and it is reported that authorities allowed him to import khat from Jijjiga, the claim that only one person chewed khat in Hargeisa cannot be accurate. Khat-chewing must have been a bit of a problem in Somaliland as early as the early 1920s for British authorities of the Protectorate who were obliged to pass a law in 1921 'forbidding the cultivation, import, and the sale of khat, except under control to meet the needs of addicts for whom it was advised that abolition should be gradual.'[8] The story nevertheless points to the limited level of khat consumption in Hargeisa before the Italian invasion.

During the Second World War, khat consumption not only increased in the border towns but also extended deep into the Protectorate including Hargeisa, despite a ban on khat importation since 1921. Two interrelated developments were critical for the expansion of khat consumption: the connection of the Protectorate with the khat production areas of the Harerge highlands by modern transportation and the migration of a considerable number of Somali pastoralists to urban centers.

By the 1930s, Somaliland had not seen much economic restructuring in the pastoralist sector after more than half a century of British colonial rule. In the inter-war years, roads were improved and maintained through the use of convict labor and paupers. The roads permitted quicker transport by trucks for Somali mutton and goat skins, which

[5] Interview: Avidas Terzian, Addis Ababa, August 8, 1994.

[6] Interview: *Balambaras* Ahmed Tewfiq Mohammed, Harer, March 23, 1994.

[7] Abdulahi Sheikh Elmi, 'Khat: History, Spreading and Problems in Somalia,' in Thomas Labahn (ed.), *Proceedings of the Second International Congress of Somali Studies* (Hamburg, 1984), p. 274.

[8] CO 323/1885/7, The Military Governor of Somaliland to Civil Affairs Branch, East Africa Command: Control of Cathe [sic] Edulis (forsk.), May 30, 1945; CO 323/1906/6, S.J. Oliver, Secretary to the Government of British Somaliland to Civil Affaris Branch, East African Command, January 17, 1947.

were particularly prized in Europe and the United States for women's gloves and luxury leather goods.[9]

Between August 1940 and March 1941, the Somaliland Protectorate fell under Italian occupation. A British coastal blockade created additional trans-frontier trade opportunities for Somali traders in the Italian occupied areas of supplying foodstuffs for the townships, which had suffered severe food shortages. The pastoralist areas faced less privation owing to a couple of years of above-average rainfall, but the herders experienced an acute shortage of clothes. Somali traders took advantage by supplying both needs, especially after several expatriate traders abandoned their shops and fled the Protectorate shortly after the Italians arrived. The Somalis' trading ventures were facilitated by the roads the Italians built, and trans-frontier trade increased dramatically, especially after the British occupation of Ethiopia.

When the Italians occupied Jijiga every Somali trader who could obtain transport rushed goods there and sold them to the fascist troops at enormous profits. Within a year, many traders were taking caravans across the border and doing very well. Traders purchased motor vehicles: the number imported into the Protectorate jumped from 6 private cars and 49 commercial trucks in 1933, to 16 cars and 230 lorries in 1937. During the same year, a total of 51 commercial and 316 private vehicles were licensed to operate, and no fewer than 600 Somalis were directly engaged in the operation as drivers and mechanics. Before 1940 most of these operations were conducted by the more experienced Indian firms, but no fewer than 150 trucks were owned by Somalis, who in many instances sold the bulk of their livestock to invest in motor vehicles. Trucks were used in running profitable companies that transported skins, live sheep and goats to Berbera and brought back mail, merchandise, foodstuffs and passengers. By 1942 itinerant Somali traders were present in virtually every village and in the vicinity of livestock markets where they bartered tea, cloth, dates, rice and sugar to the pastoralists in the Haud and the Ogaden.[10]

By 1940 Somalis who had benefitted from the brisk business were using their money to acquire plots in Hargeisa and build houses and shops. Restaurants and bars started to cater to Somalis who had taken advantage of employment opportunities and had become government clerks, interpreters, teachers and petition-writers. For young Somali males, these enterprises offered escape from the hardships of pastoral life. As pastoralists recently settled in towns, they fought the boredom of town life by socializing at chew sessions.[11] The income they obtained was partially spent on the consumption of khat. This was noted by the

[9] Samatar, *The State and Rural Transformation in Northern Somalia*, pp. 22–57.

[10] *Somaliland Protectorate Annual Report, 1936* (London, 1937), pp. 27–8.

[11] Charles L. Geshekter, 'Anti-Colonialism and Class Formation: The Eastern Horn of Africa, 1920–1950,' in Thomas Labahn (ed.), *Proceedings of the Second International Congress of Somali Studies* (Hamburg, 1984), p. 242.

Protectorate administrators when they reported that the increase in the import and consumption of khat between 1941 and 1946 was owed 'to some extent to an increase in the purchasing power of certain classes of the community.'[12] Truck drivers carrying goods and passengers to and from towns as far west as Jijjiga, near the khat-producing Harer highlands, began to rely on khat to stay awake during long trips and transported fresh sprigs on their return trip, thus making fresh khat readily available to town-dwellers.[13]

Political factors also played an important role in the expansion of consumption in the Somaliland Protectorate. As Somalis moved to towns, they formed social clubs and self-help societies, whose membership consisted of the new elite. Their goal was to promote interest in secular education, to exhort Somalis to overcome clan divisiveness in the name of Islamic unity and to seek appropriate privileges for civil servants. Over time, traders, coffee-shop owners, personal servants of British officials, truck drivers and the foreign-educated elite began to develop new interests and aspirations. They subscribed to the ideal of a united and independent Somalia. In their clubs, the intelligentsia discussed politics and began to chew khat. Young men began to participate in regular afternoon chew sessions.[14]

As the number of consumers increased, the amount of imported khat grew dramatically in a short time. In 1928 approximately 750 bundles (approximately 1 kg each) were imported into the protectorate per day and by 1936 the known amount had increased more than fivefold to 4,000.[15] In 1945 the British Consul in Harer estimated that 3,000 kilos of khat were leaving Harer every day for the Protectorate.[16] While the figure is high enough, it must be regarded as an underestimation because the amount would not be sufficient for the estimated 8,000 chewers out of the Protectorate's estimated population of 500,000.[17] Much of the demand in the Protectorate was met by smugglers, who now had the benefit of speed as against the pre-war dependence on camel transportation, but the amount of smuggled khat is unknown.[18]

After the Second World War, chewing became especially popular among small groups of poets, whose verses emphasized social solidarity,

[12] CO 323/1885/7, Governor of Somaliland to the Governor of Aden, January 21, 1946.

[13] CO 323/1906/7, Somaliland Protectorate, Government Notice No. 6, January 29, 1949.

[14] Geshekter, 'Anti-Colonialism,' p. 245.

[15] *Somaliland Protectorate Annual Report, 1930* (London, 1931), p. 17; and *Somaliland Protectorate Annual Report, 1936*, p. 13.

[16] CO 324/1924/5, Reece to Lambert, August 21, 1951.

[17] CO 323/1924/5, Officer Administering the Government of the Somaliland Protectorate to Secretary of State for the Colonies, December 13, 1951. In a petition sent to the Secretary of State for the Colonies, Somali notables state that the government's estimate was wrong. They estimate that the number of regular consumers in the Protectorate was not less than 50,000. CO 323/1924/5, Somali Notables to the Secretary of State for the Colonies, October 12, 1951.

[18] CO 323/1924/5, Officer Administering the Government of the Somaliland Protectorate to Secretary of State for the Colonies, December 13, 1951.

community of purpose and political goals. Chewing khat for hours quickly acquired the symbolic significance of solidarity and resistance in a period of heightened aspirations for self-rule and independence. The custom of chewing together from a common bundle of twigs came to represent an important ritual of friendship, mutual trust and social cohesion. Pastoralist Somalis, who previously associated the poets with Europe because of their western manners and styles, came to view them as wise counselors of the youth. In 1947, concerned about the association of chewing with politics and the effect of smuggling on the Protectorate's economy, the British Administration tried to proscribe khat importation through tighter control measures.[19] The effort backfired as more and more Somalis took up chewing as a symbol of resistance to colonial authority. The government reasoned that the prohibition was part of its program to bring about development. Many Somalis rejected it, arguing that they would rather forgo progress 'if it is going to take away from us our liberty and our harmless social habit and brings us harsh laws and racial discrimination.'[20] The law was later rescinded, but khat-chewing remained an expression of the aspiration for independence as long as the British stayed there.[21]

Urban consumption of khat continued to rise throughout the 1960s. Its trade was so lucrative that farmers who had left rural areas for urban centers began to return to the cultivation of khat from the mid-1960s onwards. As the income-earning power of khat became increasingly known in the late 1970s, even members of the urban elite, such as merchants and government officials, moved back to rural areas to profit from the new-found 'miracle' crop. Khat consumption remained widespread until 1983, when the government of Muhammad Siad Barre banned its production and import.[22]

In the period after the Second World War, the French colony of Afars and Issas or Djibouti also came to import a significant quantity of Harerge's khat. The leaf had been consumed there since the creation of the port in the late nineteenth century. The initial khat chewers in Djibouti were mostly Yemeni Arabs, who, from the founding of the port, constituted a substantial element of the city's population. They traded profitably in slaves and firearms at Obock and Tadjoura before Europeans established

[19] CO 323/1906/7, Somaliland Protectorate, The Kat Ordinance, 1947; CO 323/1906/7, Somaliland Protectorate, The Kat (Amendment) Rules, 1949; CO 323/1906/7, Somaliland Protectorate, An Ordinance Further to Amend the Kat Ordinance of 1947.

[20] CO 323/1924/5, Somali Notables to Secretary of State for the Colonies, October 12, 1951.

[21] CO 324/1924/5, Gerald Reece to Griffiths, August 2, 1951. In this letter to James Griffiths, Secretary of State for the Colonies, Sir Gerald Reece, Governor of the Somaliland Protectorate, acknowledges that there was an opposition to the government's measure expressed in the form of demonstrations and petitions, mostly in Hargeisa. He dismisses the opposition as a work of a few 'confirmed agitators looking for a stick with which to beat the Government.' Even though the governor dismisses it, the khat issue was clearly a matter of political concern he could not ignore.

[22] Abdi I. Samatar, 'The Predatory State and the Peasantry: Reflections on Rural Development Policy in Somalia,' *Africa Today*, 3rd Quarter (1985), pp. 50–2.

themselves along the Red Sea coast in the late nineteenth century. When the French transferred their capital from Obock to Djibouti in 1896, the Arab communities followed and, in a short time, accumulated large fortunes through trading and investing in urban real estate. New immigrants, mainly from Yemen, also became prosperous through hard work, thrift and self-deprivation, and added to the already numerically substantial Arab community.

For Yemeni Arabs, khat-chewing had been a longstanding and widespread practice over many centuries. When they emigrated, Yemenis took the custom to their destinations.[23] While it is difficult to find concrete evidence that Yemenis were the original khat users in Djibouti, there is ample indication that khat was the only luxury common to Yemeni property owners, shopkeepers and laborers. In the context of Djibouti's hot climate and Islamic prohibition of alcohol, most well-off Yemeni Arabs were willing to spend a considerable amount of money on this rare commodity. Over time, khat-chewing spread to Djibouti's Afars, Issas and Somali residents. In the French territory, Arabs were culturally influential, owing to the fact that Islam originated in their land. The African population viewed them as God's chosen people and their tongue as the sacred language of the Qur'an; they adopted Arab manners, one of which was khat-chewing.[24] By the 1940s most accounts claim that chewing had become a widespread ritual of the majority of the city's residents.

In the years before the completion of the Djibouti–Dire Dawa railway, Yemen probably supplied much of Djibouti's khat needs because a dhow from Aden was capable of ferrying fresh khat more quickly than camels from the Harerge highlands. After the arrival of the railway to Dire Dawa, Harer's khat became available in the Djibouti market in greater quantity. No data are available to show the number of chewers in Djibouti, but the import statistics (see Table 5.1 below) indicate that consumption had been rising steadily ever since fresh khat began to be imported from Harerge.

In the 1950s Djibouti and British Somaliland (unofficially) imported significant amounts, but Aden Colony was by far the largest importer of Harerge's khat. Since the evidence on khat consumption in southwestern Arabia is inconclusive and contradictory, it is neither possible nor necessary to try to pinpoint the beginning of chewing in Aden. Most experts believe that southwestern Arabia was where khat-chewing started; written records suggest that khat was chewed in Yemen in the fourteenth century.[25] By the

[23] Yemeni Jews who migrated to Israel after 1948 have continued to consume khat and have cultivated the tree there. Larry Derfner, 'High on Gat: Gat Chewing, an Old Yemenite Custom, is Making a Mark on Israeli Society,' *The Jerusalem Post*, International Edition (November 16, 1996), pp. 16–17.

[24] Virginia Thompson and Richard Adloff, *Djibouti and the Horn of Africa* (Stanford, 1968), p. 32. The authors provide a further description of Arabic influence among the native population of Djibouti, pp. 31–3.

[25] See, for example, Kennedy, *The Flower of Paradise*, chapter III; and Weir, *Qat Consumption and Social Change*, pp. 71–101.

sixteenth century khat-chewing seems to have become established in the highlands of Yemen, and the legitimacy of consuming the stimulant was under debate among Muslim scholars who visited or lived there.[26] The Danish explorer, Carsten Niebuhr, reported that, during his journey through Yemen in 1762, the locals were chewing khat regularly.

During the early days of British rule of Aden, there was a substantial amount of khat trade from north Yemen to Aden. In 1844 a small number of sellers were licensed by the government to bring khat from Yemen and distribute it in the port. Khat prices were exceptionally high, and often increased by as much as 1,000 percent between the farm and market, because a few merchants controlled the supply and dictated prices. The elite and lower classes of Aden consumed different varieties of khat, brought into the city from different production areas in the interior. In the late 1860s khat imports accounted for 6 percent of the total value of overland imports to Aden.[27]

There is ample evidence suggesting that khat consumption was widespread in Yemen by the early twentieth century. Immediately before the First World War, travelers reported that, with the exception of those who could not afford it, most Yemenis daily engaged in khat-chewing.[28] Amin Rihani, an American who traveled extensively through Yemen in the late 1920s, claims that khat use there was a universal practice. The Yemenis, Rahini relates, say 'ghat [khat] is our food and our coffee.' He goes on to say that they cannot go for 'a single day without ghat [khat].'[29] His statement is obviously somewhat exaggerated, but it underscores the fact that, by the time of his trip, khat-chewing was a deeply rooted cultural practice in Yemen. There is hardly any account of Yemen from the early 1930s on, written by travelers or administrators, in which khat is not depicted as a national habit of the Yemenis while also emphasizing the negative influence of khat on the people.[30]

Reports from Aden during the Second World War reveal that there was a marked increase in khat consumption in the Colony. The rise was attributed to the great increase in the population of the Colony following the massive immigration of Yemenis 'from the Yemen and Protectorate engaged in military and naval contracts during the war years.'[31] In the 1950s the port city of Aden had a cosmopolitan population of about 140,000, of which 75 percent were Arabs of Yemeni origin – individuals who had left their homelands either in search of employment or to escape the repressive government of the Imams for the relative freedom of Aden. In 1956 the Arabs, along with the Somalis, who constituted about 8 percent of the population, consumed nearly 4 million pounds of

[26] Daniel Martin Varisco, 'On the Meaning of Chewing: The Significance of Qat in the Yemen Arab Republic,' *International Journal of Middle East Studies*, 18 (1986), p. 3.

[27] R.J. Gavin, *Aden Under British Rule, 1838–1967* (London, 1975), pp. 58, 121.

[28] Wyman Bury, *Arabia Felix: Or the Turks in Yemen* (London, 1915), p. 152.

[29] Amin Rahini, *Arabian Peak and Desert: Travels in al-Yaman* (Boston, 1930), p. 46.

[30] Harold Ingrams, *Arabia and the Isles* (New York, 1966), p. 106; L. Farrago, *Arabian Antic* (New York, 1938).

[31] CO 323/1885/7, Governor Secretary of Somaliland Protectorate to Directors of Departments, Monthly Circular Letter, May 31, 1945.

khat.[32] Preferring the Ethiopian variety to the Yemeni one, these users created a large market for Harerge's khat.

Structures of khat marketing

Within Harerge, Dire Dawa was the main market center of commercial khat before Italian-built roads connected the town with the khat-producing areas. In those days, farmers are said to have harvested khat at about 6 p.m. and set out on foot the same night with head-loads or donkey-loads of khat to reach the market at Dire Dawa early in the morning. At the town's marketplace, called *chat terra*, khat was sold and then processed by the exporter's employees, who plucked off the leathery leaves, trimmed the long stems and expertly packed the good khat into handful-size bunches called *aqaara*. The bunches were then wrapped in banana leaves, which helped to retain the khat's natural moisture and freshness, and then taken to the train station for shipment to Djibouti. Farmers who reached Dire Dawa late found themselves obliged to sell their khat at very low prices to local chewers. Since khat could not be stored for the next market day, the farmers were usually left with no alternative but to sell their khat at whatever price the purchasers offered. For this reason, farmers describe the khat market before the Italians as discouraging.[33]

The khat market in Jijjiga was twice the distance to Dire Dawa from the production areas. Merchants purchased khat in the town and dispatched it by camels in two directions. The first route headed for the towns in the Ogaden and the second went farther east to border towns within the then British Somaliland.[34] Since khat was illegal in the territory, demand was met through regular smuggling from Harerge. Because there is no record of transactions conducted outside the legal realm, it is almost impossible to put figures on the amount of khat that entered Somalia through illegal channels. For the period after legalization, however, we can extrapolate from the income statistics of the Somaliland Protectorate, which, in 1959, collected an average monthly revenue of $14,000 in khat import taxes.[35] An Ethiopian government record shows that Somalia levied import taxes at a rate of US $1.00 per kilogram, suggesting that as much as 168 tons of khat was annually imported from Harerge into the Protectorate in the late 1950s.

After the war, Djibouti's khat import also increased significantly, mainly owing to the improved Harer–Dire Dawa road and the increased use of automobiles for transport. The new means of transportation enabled farmers to get khat to the Dire Dawa market and return home

[32] Weir, *Qat in Yemen*, p. 171, n. 4; Gavin, *Aden*, p. 445, App. B.

[33] Interviews: Ahmed Amit, Ganda Gole, Haramaya, May 25, 1994; Yusuf Hammo, Haramaya, May 18, 1994.

[34] Interviews: Mussa Botto, Ganda Gole, Haramaya, May 23, 1994; Rayima Sali, Ganda Gole, Haramaya, May 23, 1994.

[35] Brooke, 'Khat in the Middle East,' p. 57.

in a matter of hours and for traders to purchase fresh khat, which they dispatched by night train to Djibouti for next-morning consumption.[36] My informant, Adem Aliyu Jaylan, remembers the experience of his father in the khat trade during the late 1940s:

> By the time my father entered the khat trade, the nature of the business had changed a bit [compared to the pre-war times]. The road connecting Harer to Dire Dawa had been built and there were four passenger vehicles providing transport along the highway. Each of them left AwWoday for Dire Dawa at 6 and 9 in the morning, at 12 noon and at 3 in the afternoon. There were also three trucks, owned by Mr Gonja, Mr Georgi and the Besse company, that delivered khat to Dire Dawa. They did not receive enough freight as soon as they arrived. They drove back and forth around town and blew their horns to make people aware that they had arrived.
>
> Although the business of khat selling was not the most profitable occupation at the time, it was really taking off. The reason for its take-off is the coming of the all-weather motor road connecting Dire Dawa to Harer. The construction of the road really began to change everything here. If anything, I learned three more languages, Arabic, Somali and Amharic, in addition to my mother tongue, the Oromo language, from the people the road brought here.[37]

Indeed, khat selling was not the most profitable business in the 1940s, the period my informant describes. At the end of the decade, however, it became very profitable following the inauguration of daily air transportation between Dire Dawa and Djibouti in 1949. Thereafter, a plane daily brought 600 kg of khat to Djibouti, most of which was consumed there while small quantities were delivered to a few villages near the city that could be quickly reached by road. Within a short time, the imported amount nearly doubled and its value tripled. In 1952 imports reached approximately 21,900 kg per month and, according to government estimates, an additional 10 percent of this figure was smuggled into the country, bringing the total to approximately 24,000 kg per month. Much of this amount was consumed in the city.

Khat arrived in Djibouti in bunches weighing 280 grams each. In 1952 a total of 85,000 bunches arrived daily, corresponding to the total amount given in kilograms. Except for a slight drop in 1954, khat imports into Djibouti generally rose steadily throughout the 1950s (Table 5.1). Imports rose sharply in 1953, coinciding with increases in the total wages paid to workers. Apparently, the fluctuation in import figures was attributable to the availability of disposable cash rather than to a decline in the number of people who consumed khat. Statistics from Djibouti show a general upward swing in khat imports, reaching 410.2 tons in 1957, costing Djiboutians 328,160,000 DF or $1.5 million.[38]

[36] Interviews: Yasin Usman, Dire Dawa, February 20, 1994; Duressa Feyissa, Dire Dawa, February 24, 1994.

[37] Interview: Adem Aliyu Jaylan, AwWoday, January 19, 2000.

[38] Brooke, 'Khat in the Middle East,' p. 55.

Table 5.1 *French Somaliland (Djibouti): Khat imports,* 1952–7

Year	Volume (tons)	Value (in '000 dollars)
1952	230.4	649.7
1953	281.3	794.7
1954	263.7	805.6
1955	302.5	974.1
1956	314.2	1,180.6
1957	410.2	1,542.4

Source: Adapted from Brooke, 'Khat in the Middle East,' p. 55.

Khat was very expensive in Djibouti. This fact is more apparent when considered in the context of the income level of the port city's population and khat retail prices. The small bundles were bought from the Ethiopian producer for 20 Djibouti Francs (DF) and sold to the consumer for almost DF 200, a rise of 900 percent. Statistics from the early 1950s show that Djiboutians spent almost DF 14.5 million ($68,103) every month. This translates to an annual expenditure of DF 172,800,000 ($812,160) for a population of 18,000, including women and children. A calculation on the basis of the 1953 figure, which could be taken as an average, would indicate that the per capita monthly expenditure on khat was DF 9,600 ($45).[39] This amount of expenditure was certainly high for a country where the minimum daily wage reached DF 35 ($0.17) only in 1966, after an aggregate increase of 34 percent in eight years.[40]

In 1956 Djibouti's expenditure on khat reached twice the amount of 1953. The per capita consumption for that year was approximately 38 lb, representing an estimated retail expenditure equivalent to $87 per person. A year later these figures rose to a per capita consumption of 50 lb, at a per capita cost of $95 or $1.7 million for the total population. This was a staggering sum for a population of barely over 20,000 and whose wage levels were low. One way of getting an understanding of what the population spent on khat is to compare it with expenditure on food – DF 8,000 compared to DF 9,600 for khat. It can be estimated that a wage earner spent at least 40 percent of his income on the stimulant.[41]

Djibouti was not the only place where the khat market increased with the introduction of air transportation. With a regular air cargo service,

[39] Thompson and Adloff, *Djibouti*, p. 164.

[40] CO 1015/2051, UN Commission on Narcotic Drugs, Draft Report on the Work of the Twelfth Session, p. 4. The minimum wage was raised 5% in 1958, 8% in 1961, 6% in 1963, 5% in 1964, and 10% in 1966. Thompson and Adloff, *Djibouti*, p. 218.

[41] The value of khat was given in Djibouti Francs and was converted, where necessary, to US dollars based on the 1950s rate: DF 100 = $0.47. The figures in the text are my own calculations based on data culled from the following works: UN Commission on Narcotic Drugs, 'The Medical and Social Problems,' p. 35; Kervingant, 'The Consumption of Khat in French Somaliland,' p. 42. A 1956 survey had suggested that Issa and Afar wage earners spent about half of their monthly pay on the purchase of khat. Thompson and Adloff, *Djibouti*, p. 215.

Aden Colony became the major market for Ethiopian khat, 95 percent of which came from Harerge. Before October 1949 nearly all of the khat consumed in the Colony was imported from the Yemen Protectorate. Yemen's export to Aden had risen from 280.4 tons in 1935 to 1,097.8 tons in 1949. Perhaps owing to better taste or capacity to stimulate, air-transported Ethiopian khat quickly gained consumer preference in Aden and came to enjoy a far greater market than was ever reached by the Yemeni variety.[42] In a little more than a year after the initial shipment, Ethiopia displaced Yemen as the leading exporter, and the relative position of Yemen in the Aden market thereafter declined. In 1956, Aden imported 1,502.3 tons of Ethiopian khat at a cost of $4.6 million, as compared to 216.4 tons from Yemen valued at $793,000 or 12.5 percent of the total khat imports.

Export figures show that the amount of khat imported into Aden doubled in less than a decade (Table 5.2). Khat imports, which were 819.7 tons (all from Yemen) in 1947, reached 1,502.3 tons in 1956. The value in 1956 was twenty times that of 1947. Calculated on the basis of the city's population, estimated to be 140,000 in 1956, the per capita consumption of khat for that year was 12.2 kg, representing a per capita expenditure of $38.7, and totaling $5.4 million for the city's population. During the same years, the declared import unit value of khat increased by 900 percent, from $16.80 to $156.8. Such a sharp rise was due to the high cost of air transportation and frequently increasing taxes and duties on both the Ethiopian and Aden sides.[43]

Table 5.2 *Aden Colony: Khat imports,* 1947–57

	From Ethiopia		From Yemen		Average unit
Year	Metric tons	'000 dollars	Metric tons	'000 dollars	value in $
1947	–	–	819.7	271.6	16.8
1948	–	–	972.6	319.2	16.8
1949	NA	–	1097.8	481.6	22.4
1950	311.1	227.6	721.4	527.6	38.0
1951	1113.1	1037.4	418.0	385.8	45.3
1952	1257.0	1074.1	412.8	349.3	42.2
1953	1481.4	1925.9	545.9	694.4	64.6
1954	1494.5	3198.3	481.4	1062.0	110.3
1955	1211.6	3731.9	248.2	764.5	156.8
1956	1502.3	4627.2	216.4	792.6	156.8
1957	235.8	726.4	170.6	525.4	156.8

Source: Adapted from Brooke, 'Khat in the Middle East,' p. 56. The 1957 figures are recorded only to April 1, the date Ethiopian khat was banned in Aden Colony.

42 It is difficult to pin down the reason for the popularity of Ethiopian khat, besides speculating that preference had much to do with the stimulating quality of the chewable leaves. Khat's quality varies with different conditions of climate, soil, elevation and cultivation techniques, and it is perhaps a combination of all of these that may account for the overwhelming preference for Ethiopian khat in Aden.

43 Calculated based on Brooke, 'Khat in the Middle East,' pp. 54–5.

In response to the rapidly growing demand, Ethiopia's khat export earnings increased commensurately. We have data that show the general trend of Ethiopia's khat exports. In 1946 only 26 tons of khat was exported from Ethiopia and earned 14,000 birr ($5,644.80). A decade later, the export amounted to 1,618 tons, valued at 6,996,000 birr ($2,820,384). By 1963, the total export volume had quadrupled, reaching 4,802 tons, while earnings had risen approximately threefold. In less than two decades, khat was transformed from a crop cultivated by farmers to supplement a family's cash income to Ethiopia's fifth largest export commodity.[44]

The export earnings from khat were obviously beneficial to Ethiopia's balance of payments but, since khat was exported by private businessmen, the government's share of the khat business came from duties and taxes. Figures are hard to come by, but we can make a rough estimate of the revenue from taxes by basing our calculation on Customs revenue statistics. For most of the 1950s the Ethiopian Customs levied 1.0 birr per kilogram of exported khat.[45] At this rate, the Ethiopian Treasury had collected 26,000 birr in 1946. Just one year after air transportation started, the amount had climbed to 368,000 birr and, by 1963, it had reached 4,802,000 birr.[46]

Taxes were also collected in Harerge's various sub-provinces from the khat transactions made within the country. The aggregate figures are not available, especially for the 1940s, when the Ethiopian government was not organized enough to keep records. Fortunately, we have figures showing the amount of taxes collected between 1948 and 1952 Ethiopian Calendar (1955/6–1959/60). The figures show tremendous fluctuation, perhaps reflecting corruption or poor bookkeeping, but they do represent a rough estimate of the amount of the tax revenue from khat. During the five years for which we have records, the government collected a total of 12.3 million birr, or an average annual revenue of 2.5 million birr.[47]

In 1962 the combined total of export duties and taxes on domestic transactions amounted to more than 7 million birr or $3 million. The revenue from khat was three times the amount of the total land tax collected in 1962 for the entire Harerge province, which was 2.4 million birr.[48] Ethiopia's total revenue for the same year was 218.8 million birr. It is therefore not difficult to imagine the political implications of the

[44] Amare and Krikorian, 'Coffee's Rival,' p. 370.

[45] HPA-KF.1, Negatu Mekuria, Harerge's *Bejirond* to *Dej.* Workineh Wolde Amanuel, Harerge's *Inderassie*, Harer, September 16, 1960.

[46] See Appendix A for the export quantities. The estimates are based on calculation of the export quantity and the 1 birr duty per kg of khat.

[47] HPA-KF.1, Khat Tax Bureau to Harerge Provincial Administration Office, Report on the revenue collected from khat between 1955 and July 6, 1960, Harer, August 28, 1960. Another document shows that 855,423.50 birr was collected from July to October of FY 1952 E.C. It is fair to assume that an average of 2.5 birr million was collected in tax revenue annually in the 1950s, if the mentioned amount was collected in four months. HPA-KF.1, Harerge's *Bejirond* to the Minister of Finance, Mahteme Sellassie Wolde Masqal, December 4, 1960.

[48] Seleshi Wolde-Tsadik, *Land Taxation in Harerge Province*, Imperial Ethiopian College of Agriculture, Experiment Station Bulletin 48 (Dire Dawa, 1966), p. 11.

dramatic economic gain for the exporting side and the potential problem for the importing side.

Government reactions

From the standpoint of the Aden and Djibouti governments the rapid growth of khat consumption and trade was socially and economically dangerous. During the early 1950s, the Aden Colony government's concern about the rapid increase of khat imports and its implications for the Colony's balance of payments was evident in the editorials of Adeni newspapers and in Legislative Council debates.[49] Initially the government attempted to limit imports by imposing a heavy duty and a high local tax, as ways of discouraging imports without causing social convulsion. The effort brought in large revenues for the government, but did not curb the rise in consumption. As we have seen, high taxes and duties on khat succeeded not in reducing imports, but rather in raising consumer prices. The continued sterling flow from Aden to Ethiopia, therefore, remained a matter of serious concern for the Colony's government and merchants.[50] Pressed by constituencies, the Aden Legislative Council requested the government to impose restrictions on khat imports from Ethiopia. Accordingly, importation of Ethiopian khat was banned as of April 1, 1957. Health reasons were cited to justify the ban, but contemporary sources agree that the main reason was to prevent hard currency from draining into Ethiopia.[51]

The Ethiopian government considered the loss of revenue so serious that it took dramatic measures to force the Aden government to reverse its decision. One such measure was the revoking of Aden Airway's license to operate its commercial flights to Addis Ababa. The Aden government did not flinch for more than a year, even though it was evident that its ban was ineffectual. Khat importers and retailers, estimated to have numbered 51 and 190 respectively, re-established their trade at Dar Saad, a village located right on the Colony's border with the Sultanate of Lahej, and had no difficulty arranging increased supplies from Yemen and Ethiopia. The Ethiopian leaf was preferred to the Yemeni variety and the demand for it was met by imports from Djibouti and the Assab port of Ethiopia, whence the freight was ferried by speedy boats to the coastal villages of Lahej and then by trucks to

[49] Aden Colony, *Proceedings of the Legislative Council's 11th Session*, Printed by the Government Printer, Aden, February 28, 1957; FO 371/126868, Simmonds to Morgan, November 15, 1957.

[50] CO 1015/2050, Mohamed Hason Khalifa, 'Qat: The Enemy of Society,' undated. This document puts the annual currency outflow at 50,000,000 shillings. This figure roughly corresponds to the 4,000,000 shillings spent on khat monthly, quoted in Legislative Council Debates, Aden Colony, *Proceedings of 11th Session*, p. 8.

[51] CO 1015/2050, Luce to Furlonge, April 5, 1957; SD 875.00/7-957, American Embassy to State Department (SD), 'Weekly Economic Review,' April 15, 1957.

Dar Saad.[52] Soon after the ban went into effect on April 1, 1957, two to three thousand Aden residents began to descend on Dar Saad each evening, with double that number on weekends.[53] Only the Aden government seemed to be affected by the ban, sustaining a revenue loss from taxes and duties on the leaf amounting to 50,000 pounds annually.[54] It also became evident that the nightly gathering of thousands of khat chewers with a grievance against the government would sooner or later present a grave security problem.[55] A Commission of Inquiry was called in February 1958 to study the matter. Its final report confirmed the conventional wisdom that the ban was not working and recommended its repeal. Consequently, the government rescinded the import ban on June 24, 1958 and, in its place, instituted a system of licensing as a measure of control over imports and prices.[56]

The effects of Aden's short-lived ban on Ethiopian khat were mitigated by the action taken by another British dependency, the Somaliland Protectorate. Before the 1950s, the British administration of the Protectorate fought against khat smuggling vigorously. Smugglers were jailed, even shot, and their trucks tossed into the sea, but none of those actions stopped the illegal flow of khat into the territory.[57] Banning khat was evidently unworkable and counterproductive. Shortly after khat was prohibited in Aden, the Somaliland Protectorate removed a long-standing ban on the leaf. A decree was promulgated to repeal all of the khat prohibition laws and provided for importing only by surface conveyance. Importation by air was proscribed in order to keep the market at a reasonably small level. The restriction did not have much effect because most of the khat market centers in the Protectorate were only a short distance away from the production areas of the Harerge highlands.[58]

The French government encountered problems in Djibouti similar to those of Aden in regard to khat imports and consumption. As in Aden, several measures were taken progressively to reduce the level of imports and consumption but none proved successful. The Djibouti administration imposed taxes on khat sales for the first time in 1952 in an effort to discourage merchants from engaging in khat business. Contrary to the anticipated effect, official imports held steady and consumption tripled in three years, thereafter showing growth with increases in the minimum wage. Although successive governors inveighed against the evils of khat and took measures to limit importation or consumption, their efforts only increased smuggling and the loss of government revenue. The

52 CO 1015/2050, Simmonds to Morgan, November 15, 1957.

53 CO 1015/2050, Simmonds to Morgan, April 27, 1957.

54 CO 1015/2050, Luce to Furlonge, April 5, 1957.

55 CO 1015/2050, Simmonds to Morgan, July 27, 1957.

56 Aden Colony, *Report of the Qat Commission of Inquiry* (Aden, 1958), pp. 16–17.

57 CO 1015/2050, Stebbing to Hall, February 13, 1958; CO 1015/2050, Extract from 'War Somali Sihidi' No. 111, April 6, 1957.

58 CO 1015/2050, Pike to Secretary of State for the Colonies, April 6, 1957; Elmi, 'The Chewing of Khat in Somalia,' pp. 166–7.

governors were unable to take more effective measures, fearing that prohibition would alienate the Ethiopian government and result in reduction of revenue for the Djibouti government. Only smugglers stood to profit from a ban on khat imports. The territorial assemblymen were reluctant to take any legislative action because they did not want to risk their political future by voting to forbid the importation of khat.[59]

When it became clear that the colony government was unable to take any measure to halt the increasing imports and consumption of khat, the French government felt obliged to step in and do something. Accordingly, Paris placed khat on its list of narcotics in 1957, and proscribed the import, export, production, possession, trade and use of it in any form in French-administered territories,[60] despite the World Health Organization's declaration that khat did not fall within the standard definitions of addiction-producing drugs. The Ethiopian government made its displeasure known by declaring its view that khat was not in any sense a narcotic drug that should be purged from international trade. Aware of Djibouti's economic dependence on Ethiopia, for which it was the chief port and eastern terminus of the Franco-Ethiopian Railway, the Governor-General of the French Territory of Afars and Issas (FTAI) resisted the implementation of the decree in his territory. The decree was thus executed into law in metropolitan France, but khat imports and consumption in Djibouti continued unabated.[61]

The Ethiopian government opposed the Aden and Djibouti governments' attempt to prohibit the importation of khat for good reasons. In addition to the revenue that accrued to its treasury, khat was a profitable freight for Ethiopian Airlines. In 1957 khat represented about 12 percent of its total revenues – about $45,000 per month.[62] As an important source of income for farmers, khat was also viewed as a politically stabilizing commodity. This view is expressed in several administrative letters and documents, in which government officials speculated that the loss of khat revenue might lead to discontent and provoke rebellion.[63] Furthermore, the Ethiopian government used the revenue from khat to fund development programs and security efforts in Harerge's politically troubled lowland regions, the Ogaden, the Haud and the Reserved Areas.

Revenues, politics & conflicts

Predominantly inhabited by Somalis, the Ogaden and the Haud were transferred from British to Ethiopian rule in 1948 and 1954, respectively. The Ethiopian government did very little to integrate the regions

[59] Thompson and Adloff, *Djibouti*, p. 164.
[60] *Journal Officiel*, 5 Avril, 1957.
[61] Brooke, 'Khat in the Middle East,' p. 56.
[62] SD 875.00/4-557, Embassy to SD, 'Weekly Economic Review,' April 5, 1957.
[63] HPA-GC.1, Tamirat Yigezu to Haile Sellassie, June 30, 1964; HPA-GC.1, Tamirat Yigezu to Yilma Deressa, June 19, 1964.

economically, although it did coopt a few Somali leaders into the national governing structure. The isolation created a favorable condition for Somali nationalists to incite Ogaden residents against the Ethiopian government through broadcasts from Mogadishu and Hargeisa and by using Somali pastoralists who had grazing rights within Ethiopian territory according to the 1897 and 1954 treaties between Ethiopia and Britain.[64]

Several skirmishes between Ethiopian and Somali forces demonstrated that economic integration was more important than token political inclusion. The Ethiopian government consequently became inclined toward investing in the economically neglected regions. However, the tax base in Harerge was insufficient for the realization of development projects in the Ogaden. The only available funds derived from the tax revenue raised by the Khat Tax Collection Agency, which operated under the direct auspices of the Central Treasury. Regional administrators did not have jurisdiction over the agency's resources since all revenues were transferred to Addis Ababa.[65] Any withdrawal, even if borrowed by the regional administration to cover occasional financial shortfalls, had to be authorized by Emperor Haile Sellassie.[66]

A number of documents show that the emperor used khat revenues to finance the integration of the Ogaden and the Haud into the administrative structure of Ethiopia.[67] In 1959 the Ogaden received 2,033,685 birr of khat money for administrative, security and development expenses. Khat revenues were used for the annual budgets of Jijjiga, Kebri Dehar and Adal-Issa sub-provinces, for construction of office buildings for such administrative institutions as court houses, police headquarters, the provincial governor's mansion and customs offices all over the Ogaden as well as for building schools, health dispensaries and roads.[68] The start-up capital for

[64] HPA-GC.1, Aemero Sellassie Abebe, Deputy *Inderassie* of Harerge, to General Abiy Abebe, Minister of Interior, February 23, 1961; HPA-GC.1, Capt. M.G. Alexander, IESF Advisor, to Deputy Governor of Harer, January 31, 1961.

[65] HPA-KF.1, Aemero Sellassie Abebe, Deputy *Inderassie* of Harerge, to Zenebe Haile, Vice Minister of Finance, August 30, 1960. The *Inderassie* argues that according to *Negarit Gazetta*, First Year, No. 1, he, not the Central Treasury, should be in charge of the khat revenue in his province. Kifle Irgetu, shortly after his appointment as *Inderassie* of Harer, addresses the same issue in a letter to Haile Sellassie. HPA-KF.1, *Dej.* Kifle Irgetu to Haile Sellassie, September 23, 1960.

[66] HPA-KF.1, Harerge's *Bejirond* to Minister of Finance, Mahteme Sellassie Wolde Mesqel, December 1, 1960. The *Bejirond* explains that spending has always outrun revenue and requests the Minister to seek authorization for 19,000 birr from khat revenue to make up for a shortfall faced by the regional administration. HPA-KF.1. *Dej.* Kifle Irgetu to HIM Haile Sellassie, September 7, 1960.

[67] Rumors are rampant that the Duke of Harer, who was reportedly more involved in business or accumulation than other members of the royal family, had extensive financial interests in khat. Indeed, there are allusions in contemporary documents to the royal family's involvement in khat business but no concrete substantiation. SD 875.00/7-2456, Embassy to SD, Confidential Annex to Economic Review of the Year 1956.

[68] HPA-KF.1, Auditor's Report to the Ministry of Finance on the Khat Revenue and Administrative Expenditure, 1951–2. October 10, 1960.; HPA-KF.1, Aemero Sellassie Abebe, Deputy *Inderassie* of Harerge to *Balam.* Mahteme Sellassie Wolde Masqal, June 24, 1959.

establishments such as the Ogaden Water Works Development Corporation and the Lowlands Road Construction Agency was also obtained from khat revenue.[69]

Such was the significance of the khat business that Addis Ababa foresaw that domestic revenue would be augmented if the government gained greater control over the organization and management of its export. No concrete measures were taken to take over the khat business, although those implemented were clearly intended gradually to assert control over khat export, particularly to Aden.

Until 1960 khat was exported to Aden by individual businessmen operating private firms under licenses from the Dire Dawa Municipality. Some of the private khat exporters were local Somali and Oromo traders, but most were Yemeni Arabs. Over the course of time the Arabs, long involved in Harer's import–export trade, became the principal khat exporters to Aden.[70] As the khat business increased in the 1950s, more and more Adeni merchants came to Dire Dawa to take part in the growing trade, leaving their relatives to run the operations on the Aden end. Kinship played an important role on the exporting side, giving a competitive edge to the local exporters in regard to purchasing good-quality khat. This engendered a competition between the local and expatriate exporters and eventually produced an atmosphere of intense rivalry in which the khat business was conducted.[71]

Determined to win, the expatriates invited non-khat traders to join them and establish a khat-exporting company with the aim of monopolizing the trade and forcing out their competitors. With the assistance of a few local associates, the expatriates announced the establishment in 1960 of a company, the Chat Exporting Company of Dire Dawa (CECDD), with a capital of 18,000 birr. Customarily, licenses were issued by the Dire Dawa Municipality, but, in the case of the new company, the issuer was the Ministry of Trade and Tourism in Addis Ababa, indicating the central government's growing interest in the khat trade.[72]

CECDD was essentially a private enterprise, owned and operated by shareholders,[73] and expected to work in partnership with an importing agency, the Qat Importing Company Limited (CHATCO), based in Aden. The Aden government granted the latter the exclusive license of importation and wholesale distribution to retailers. In fact, since CHATCO was owned and managed by the relatives of the Arabs who had established CECDD in Dire Dawa, it may well be described as a

[69] HPA-KF.1, Aemero Sellassie Abebe, Deputy *Inderassie* of Harer to Zenebe Haile, Vice Minister of Finance. September 8, 1960.

[70] Interview: Hajji Mohammed Abdurahman, Dire Dawa, March 26, 1994.

[71] Interview: Hajji Dawd Abdo, Dire Dawa, March 26, 1994.

[72] The officials of the company included local traders such as Abdu Selam Hajji Musa (deputy general manager) and Arabs named Mohammed Hussien al-Besiri, Hussien Rubah (president), Said Besiri (accountant). Interview: Hajji Mohammed Abdurahman, Dire Dawa, March 26, 1994.

[73] HPA-KF.1, 'The by-laws of the Khat Exporting Company of Ethiopia,' 1960.

subsidiary of the latter.[74] For business purposes, however, the two entities operated as though they were separate private companies.[75]

The Ethiopian government granted the CECDD a license as the sole khat exporter to Aden, hoping to gain control over the quality of the exported leaf and to prevent a recurrence of the alleged sanitation deficiencies that had led to the 1957 ban. For about a year the company enjoyed near-monopoly rights over khat export to Aden. Meanwhile, conflict arose between the company's shareholders and the private exporters who were left out of the franchise. Ethiopian exporters blamed the 'Arabs,' who constituted the majority of the company's shareholders and managers, for allegedly attempting to sabotage the trade. Farmers accused the company of offering depressed prices, degrading the quality of khat and coercing them into selling at lower prices.[76] The situation is said to have harmed farmers and the khat business in general to the extent that candidates for parliamentary seats during the election of 1960 used the 'khat problem' as a campaign issue by promising farmers to end their predicament.[77]

Police and intelligence reports reveal that the imperial government was concerned about the political overtones of the 'khat problem' and that officials felt the issue needed resolution before it escalated.[78] One year after the establishment of the CECDD, the emperor instructed the then Minister of Trade and Tourism, Yilma Deressa, to replace the leadership of the company with 'people whose loyalty to the emperor and the country was unquestionable.' Subsequently, the minister announced the dissolution of CECDD and the establishment of a new company to 'save the work from total break-down.'[79] An Armenian called Katching Boghossian was appointed as general manager and a certain Yirgashewa Meteleya as his deputy. The new appointees took over the assets and workers of the defunct company, now named the Khat Exporters Company of Harer (CHATHA).[80] The action was hailed by local exporters and government

[74] HPA-KF.2, Head of Criminal Investigation and Intelligence Unit of Harerge Police to the *Inderassie*, November 8, 1960.

[75] HPA-KF.1, Ali Rabi, Managing Director of CHATCO, to the Managing Director of Khat Exporting Company of Harar (CHATHA), 'View Points of the Board of Directors of CHATCO on the Present Situation,' July 25, 1962.

[76] HPA-KF.2, *Graz*. Bizuneh Wolde Mariam, Governor of Gara Mula'ata sub-province, to the Provincial Administration Office, October 15, 1961; HPA-KF.2, Lt Colonel Bereket Gebre Medhin, Police Commandant of Harerge Province, to the *Inderassie*, July 17, 1960.

[77] HPA-KF.2, Lt Colonel Seyoum Gessesse, Police Commandant, to *Dej*. Kifle Irgetu, *Inderassie* of Harerge, May 8, 1960.

[78] HPA-KF.2, Brigadier General Mekonnen Deneke, Minister of State for Public Security, to *Dej*. Kifle Irgetu, *Inderassie* of Harerge, August 8, 1960.

[79] FO 371/146597, D.C. Lees, British Consul in Harar, 'Memorandum, The Ethiopian Qat Exporting Company,' Dire Dawa, August 20, 1960.

[80] *Ethiopia Dimts* (The Voice of Ethiopia), Amharic daily, October 5, 1961; HPA-GC.1, Appeal to the *Inderassie* of Harerge, signed by 49 shareholders of CECDD, October 17, 1961.

officials as 'wresting the business from the Arabs who were intent on undermining Ethiopia's economy.'[81]

The new company differed from its predecessor in several important ways. It was established by an edict, run by government-appointed managers and supervised by the Ministry of Trade and Tourism. It was essentially a government-run enterprise. Unlike the CECDD, which operated without a written contract with the importing company, the new company signed an agreement with the importing agency in Aden to deliver a daily minimum of 3,000 kg of khat during most of a year and 6,200 kg during the rainy season when khat became available in large quantities. The writing of a contract was an indication that the relations between the two companies were not based on kinship, trust or understanding, as was the case with the previous one, but strictly on business.[82]

Meanwhile, the business of exporting khat to Djibouti and northern Somalia remained in private hands. The operations of the private firms were quite distinct from the companies, especially in their purchasing mechanisms. Agents purchased khat in village markets, alongside the Harer–Dire Dawa highway, and on individual farms.[83] In the course of time, these agents developed efficient relations with farmers, which enabled them to purchase khat futures. The producers preferred this type of arrangement because it provided decent prices and eliminated uncertainties at the Dire Dawa market. Farmers traveled to the company's purchasing center not just for the prices it offered but because the Djibouti and Somalia markets did not have the capacity to absorb the entire available supply.[84]

Despite its efforts, the company's operations remained ineffectual. CHATHA failed to meet the quality and quantity demanded of CHATCO, giving rise to mistrust between the companies. Correspondence reveals that CHATHA was too haphazard in the way it was organized to fulfill its contractual commitments. It did not have purchasing stations close enough to the producing areas to buy sufficient khat to meet Aden's demand.[85] CHATHA regularly failed to dispatch the daily minimum of 3,000 kg and the brand preferred in Aden. During the months of June and July 1963, according to the agreement, 6,200 kg of khat had to be sent to Aden, but CHATHA dispatched an average of 3,000 kg. There was no day when CHATHA shipped a consignment whose weight met the agreed-upon

[81] HPA-KF.2, Second Lieutenant Tsegaye Azene, Dire Dawa Police Detective, to Police Commandant of Dire Dawa sub-province, December 8, 1960.

[82] HPA-GC.1, Agreement between Qat Importing Co. of Aden and Khat Exporting Co. of Harar, October 27, 1962.

[83] HPA-GC.1, Katching Boghossian, General Manager of CHATHA, to General Manager of CHATCO, Aden, August 1, 1965.

[84] Interviews: Amin Usman Mohammed, Ganda Issa, Haramaya, May 17, 1994; Mume Ahmed, Faldhaana, Kombolcha, June 8, 1994.

[85] HPA-GC.1, Ali Rabi, Managing Director of CHATCO, to the Governor-General of Harer Province, April 8, 1963.

amount.[86] CHATCO protested that the terms of the contract were violated by CHATHA's failure to ship the agreed-upon quantity but also by its exporting inferior-quality khat. The latter responded that its failure was owing to lack of rain, which should be regarded as an act of God.[87] CHATCO was well aware that Djibouti and Hargeisa were supplied with the best khat from Harerge and could not accept CHATHA's claim of *force majeure*.[88] On February 13, 1962, it refused khat worth Shs 49,855 from CHATHA, because the produce was declared 'unfit for human consumption.'[89] On the next day, despite CHATCO's instruction that no khat be shipped to Aden without further notice, CHATHA dispatched khat worth Shs 33,097.20, which passed inspection. CHATCO paid for it, but deducted the air-freight charges for the condemned khat. In the process, CHATHA lost close to Shs 50,000.[90]

Negotiations began in Aden between Ali Rabi of CHATCO and Katching Boghossian of CHATHA. With assistance from high-level Ethiopian government officials, the two men hammered out an agreement wherein CHATHA agreed to improve the quality and quantity of its exported khat. To that end CHATHA agreed to establish purchasing sites in the towns of Harer, Haramaya, Langie and Chelenqo,[91] but these did not help CHATHA to overcome its ineffectiveness. Business relations further deteriorated when CHATHA repeatedly dispatched delayed khat on Fridays, which was a holiday in Aden.[92]

Meanwhile, the Aden press was calling for a ban on Ethiopian khat. By 1963 the Pan-Arabist ideology championed by Egypt's Gamal Abdel Nasser was taking root in the region, and newspapers and other mass media reflected its resonance in Aden. Ethiopia's alleged friendship with

[86] HPA-GC.1, Ali Rabi to the Manager of CHATHA, 'On the lack of adaptation to agreement,' May 1, 1962; HPA-GC.1, Statement showing quantities of khat demanded by CHATCO, minimum provided by agreement, quantities of khat exported by CHATHA, May 1 to August 10, 1962.

[87] HPA-GC.1, K. Boghossian and Teferi Lemma to Mustafa Refat and Saeed Haitham, Representatives of CHATCO, August 28, 1962.

[88] HPA-GC.1, Ali Rabi to the Managing Director of CHATHA, 'View Points of the Board of Directors of CHATCO on the Present Situation,' July 25, 1962; HPA-GC.1, Mustafa Refat and Saeed Haitham, Accredited Representative of CHATCO to Deputy Governor of Harer, September 4, 1962; HPA-GC.1, 'Comparative statement to show the quantities of Ethiopian khat exported to Djibouti by Individual Importers [sic] during CHATHA's alleged *force majeure*,' enclosure item 26, July 2, 1962.

[89] HPA-GC.1, telegram from CHATCO to CHATHA, February 13, 1962; HPA.GC-1, H.R. Robertson to CHATCO, Medical Certificate, Aden Municipality, February 13, 1962.

[90] HPA-GC.1 C-1, H.R. Robertson, Officer of Health of Aden Municipality, to CHATCO, Certificate of Health, February 14, 1962; HPA-GC.1, Abdulkader Yassin Kaid, Agent for CHATHA, to General Manager of CHATCO, March 11, 1962.

[91] HPA-GC.1, Ali Rabi, Managing Director of CHATCO, to the Governor-General of Harar Province, Aden, April 8, 1963.

[92] HPA-GC.1, Negatu Mekuria, Deputy General Manager of CHATHA, to *Qegnaz.* Demissie Teferra, deputy *Inderassie* of Harerge, December 1, 1962; HPA.GC-1, Ali Rabi, Managing Director of CHATCO, to the General Manager of CHATHA, Aden, July 14, 1963.

Israel and its continued ability to drain away Aden's wealth became an issue of national debate. The print media and radio programs pressed hard for substitution of Ethiopian khat by imports from the 'brotherly neighbor' country of Yemen.[93]

During negotiations, CHATCO threatened to stop importing khat unless the Ethiopians agreed to import fish and salt in exchange. The Ethiopians rejected the condition, claiming that there was no market for those commodities within their country. The two companies' failure to strike a deal, coupled with a resentful Adeni public opinion, forced CHATCO to cease importing Ethiopian khat on June 1, 1963.[94]

In October of the same year Ethiopia's khat business suffered another blow. Sensing that Ethiopia was politically weak (as manifested in the attempted coup d'etat of December 1960) and militarily unprepared (in view of the fact that an Ethiopian contingent was on a United Nations assignment in the Congo), Somalia's leaders decided that the time was ripe for recovering their land and liberating their kin from Ethiopian domination. In 1963 they sent forces into Ethiopian territory. The hostilities that ensued disrupted transportation and even hampered trading activities in the Ogaden and the Haud. In 1964 Ethiopia declared a State of Emergency and banned the export of khat into Somalia, hoping that shortages would lead to internal discontent and instability. Only Djibouti was left to import Ethiopian khat. The termination of official exports to Aden and northern Somalia marked the end of the first phase of expansion of the production and marketing of khat.[95]

Summary & conclusions

While khat-chewing has long had a symbolic meaning for Harerge's religious and cultural elites, that meaning changed as new social groups took up the habit. This must be examined in the context of overall socio-economic and cultural developments in the Horn of Africa, including the integration of the societies in the region into the modern cash economy. It is nonetheless important to note that chewing began among different societies for different reasons: for some, chewing was the only available pastime (Christian Ethiopians); for others, it was an expression of nationalism and political protest (Somalis); and, for still others, it was a long-standing cultural ritual (Yemeni).

[93] *Al-Qalam Al-Adani,* A motion by a municipal councillor, February 13, 1963, no. 477; *Al-Ayam,* Editorial, no. 683, October 26, 1960. HPA-GC.1, Tilahun Abebe, Director of Dire Dawa Customs, to Lt Colonel Solomon Kedir, Assistant Minister for Ethiopian Customs, Report on Aden Radio broadcast on the khat situation, June 10, 1964.

[94] HPA-GC.1, *Qegnaz.* Bekure Wolde Giorigis, Governor of Dire Dawa sub-province, to Harerge Provincial Administration Office, June 12, 1964.

[95] HPA-GC.2, *Fit.* Tewahade Woldeyes, Governor of Harer sub-province, to Harerge Provincial Administration Office, October 29, 1964.

That khat was a quickly perishable commodity was a decisive factor in the process of the shrub's commodification. It determined price, making the leaves so expensive that average consumers in Djibouti spent as much as 40 percent of their incomes on khat. The fact that importing countries had to expend a significant amount of capital on khat affected their balance of payments and led to initiatives to ban the trade. For the exporting country, the khat trade generated revenues that could be used for accomplishing important political goals. Attempts at restricting consumption led to conflicts among governments, between consumers and their governments and between traders who sought to monopolize the trade and those who wanted to get a share of the lucrative business. The policies of restricting consumption, for balance of payments purposes or for other political reasons, eventually led to a permanent ban on importation, in the case of Aden, and a suspension, in the case of northern Somalia.

Six

Export Trade
Constraints, control & contraband
1964–74

> We were not responsible for creating illegal trade. We continued to do what we have done since 'the creation of the earth' [for a long time]. But the authorities said it is illegal. It seems they do not realize that there is no way a man can control another man completely.
>
> – Hajji Dawd Abdo, 1994

Growth of Harerge's khat production, which had been strong in the 1950s and early 1960s, slowed considerably after 1964. The industry faced a series of problems stemming from the sudden demise of the Aden market and the trade embargo imposed in the aftermath of the Ethio-Somali conflict of 1964. Reacting to diminishing demands and falling prices, farmers slowed cultivation of khat and, in effect, suspended the two-decade-old process of replacing Harerge's longstanding sorghum–coffee cropping complex with khat plantations. Exporters and domestic dealers scrambled to open new markets, but did not find a viable replacement for khat in their economic life. For the government, the declining khat market meant a substantial loss of revenue and a diminished capacity to deal with existing security problems and the potential political unrest arising from economic dislocation.[1]

The disappearance of foreign markets and the consequent loss of incomes and revenue set the stage for competition over the infrastructure of khat production and exchange. Focusing on the period between the declaration of a State of Emergency (SOE) in Harerge province in 1964 and the overthrow of Haile Sellassie's government in 1974, this chapter examines the reaction of farmers, traders and the government to and the implications of a relatively smaller khat industry for the region's political economy. It opens with the state of affairs just after the 1964 Ethio-

[1] HPA-GC.1, Lt Colonel Tamirat Yigezu, *Inderassie* of Harerge, to HIM Haile Sellassie, June 30, 1964; HPA-GC.1, Lt Colonel Tamirat Yigezu, *Inderassie* of Harerge, to Minister of Finance Yilma Deressa, June 19, 1964.

Somali skirmishes, examines traders' responses to increased state intervention in the khat economy, and concludes with a discussion of the political and economic implications of increased smuggling.

These apparently disparate analyses are connected. In the mid-1960s, prompted by political considerations, security concerns and the need to raise revenue, the Ethiopian government imposed restrictions on the khat trade and raised taxes. Every measure the government took distorted the marketing process, blurred the distinction between legal and illegal transactions and gave rise to a structure of distribution that bypassed or paralleled official marketing channels. The expansion of khat distribution through unofficial channels, ironically, was a direct consequence of government actions.

The aftermath of the Ethio-Somali hostilities of 1964

Before 1964 the Aden market alone constituted a 20-million birr industry. Ethiopian Airlines, the Khat Exporters Company of Harer and the Ethiopian Customs each collected annually approximately 5 million birr from exporting khat to Aden, while the same amount was reportedly shared by Harerge's farmers and traders. Estimates indicate that the government sustained an annual revenue loss of some 2.7 million birr in duties from khat export to northern Somalia[2] and another 2.4 million birr in sales taxes from domestic transactions.[3] Under the terms of the SOE, a few traders were permitted daily to supply consumers in Jijjiga with 200 kg of khat, but all traders were barred from marketing khat in areas east of the town. For farmers, the loss of income was compounded by a drought and pest infestation, which destroyed other crops in 1964[4] and diminished land-use tax revenue for the government.[5]

The demise of foreign markets was obviously a serious blow to Harerge's economy, but its consequences were not only economic. Provincial officials feared that the economic dislocation caused by the loss of income would trigger rural unrest at a time when the government's capacity to respond was seriously constrained. In an effort to

[2] This figure is based on reports that 7,000 to 8,000 kg of khat was smuggled daily into Somalia. Since customs duties were at a rate of 1 birr per kg of khat at the time, we may extrapolate the revenue loss for the year.

[3] HPA-GC.2, Governor of Jijjiga and Colonel Yilma Teshome, Chief Administrator of the SOE for the northern part of Harerge, to Major General Abebe Gemeda, *Dej.* Workineh Wolde Amanuel and *Bejirond* Negatu Mekuria, October 18, 1969. The letter indicates that the Internal Revenue Department used to collect 6,000 to 7,000 birr in income taxes daily. The figure in the text is a calculation based on the average figure of 6,500 birr.

[4] Miller and Telahun, *Organization and Operation*, p. 19.

[5] HPA-GC.2, Lt General Kebede Gebre, *Inderassie* of Harerge, to the Commander of the Third Command and Chief Administrator of the SOE, March 7, 1963.

avert what they perceived to be a potential danger, provincial officials urged the national government to try to reopen the Aden market, but its efforts failed. Aden's decision to ban Ethiopian khat was based largely on the political and economic concerns of the Aden government rather than on the organizational ineffectiveness of CHATHA claimed by the Adeni importers. Thus the Adeni market was permanently closed, leaving khat traders to look elsewhere for business.

Even when Aden was an active importer, it was northern Somalia that took most of the khat from Gursum, Harer Zuria and Wobberra *awrajas*.[6] In the years before the Ethio-Somali hostilities, Hargeisa and the Ogaden lowlands consumed approximately 7,000 kg daily. This movement was discontinued when the SOE proscribed all kinds of commercial or traditional exchanges and barred traders from operating in the area east of Jijiga and the Ogaden country. Only Djibouti continued daily to import 6,000 to 7,000 kg of khat, but the amount was not sufficient to absorb the surplus created by the disappearance of the other two foreign markets.

Reopening the Hargeisa and Ogaden markets was critical for the economic stability of the region. Farmers and officials, especially of those regions whose economic well-being depended on a flourishing Hargeisa market, submitted petitions to the provincial administration, requesting the lifting of sanctions so as to facilitate the unencumbered flow of khat.[7] Nevertheless, the petitions were often overlooked because political considerations appeared to take precedence over economic gains. The SOE was intended to curb subversive political activities, foster social discontent within Somalia by creating shortages of consumer necessities and prevent the Mogadishu regime from launching another attack on Ethiopia. It was particularly felt that withholding khat from Somalis might serve to incite them against their government.[8] For political reasons, therefore, no effort was made immediately to reopen the northern Somalia market.

It did not take too long before it became clear that the embargo would not achieve its political objective. Khat was not an ordinary commodity that consumers used when it was available and let go when it was hard to obtain. Scientific research on the physiological and psychological effects of moderate khat-chewing has turned up very little evidence implicating the leaf in serious harmful effects, but observations indicate that khat possesses properties akin to most psychoactive substances; its users become psychologically dependent to the extent that

[6] HPA-GC.2, Gugsa Abegaz, Director of Harer Zuria sub-province and the Municipality to the Office of the Provincial Administration, March 1, 1965. The Director states that his office was overwhelmed by petitions from farmers for the opening of the Hargeisa route. There are numerous such appeals in my own possession.

[7] HPA-GC.2, *Fit.* Tewahade Woldeyes, Governor of Harer Zuria sub-province to Office of Provincial Administration, October 29, 1964.

[8] HPA-GC.2, Brigadier General Abebe Gemeda, Commander of the Third Infantry Division and Chief Administrator of the SOE to Lt Colonel Tamirat Yigezu, *Inderassie* of Harerge, December 21, 1964. This is a strategic document in which the General evaluated the state of the SOE.

their physical and social functioning becomes impaired when the stimulant is not taken. The many attempts to ban khat have shown that it is impossible to prevent users from going to great lengths and paying higher prices to obtain the substance. The prohibitions imposed by the SOE, and the demand generated, gave rise to a situation favorable for dealing in khat, albeit unlawfully, in the restricted areas. The business involved the risk of being caught and punished for selling illegally, the costs of avoiding detection and the low margin of profit of trading in smaller quantities. But prices were probably higher than for khat sold in the official market, thus encouraging traders to smuggle khat to the restricted areas.

The smugglers used three major routes. Originating in Jijjiga smugglers headed northwest to Djibouti, then veered southeast into Somalia. Others took a southeasterly direction from Jijjiga, appearing to head in the direction of the Ogaden country, but quickly turned northeast to reach Hargeisa. Border guards usually concentrated on the most frequented crossing points but were unable to control the entire border. Owing to the increased risk involved in smuggling, khat may have become expensive, but there is no evidence that it was in short supply in Somalia during the time the SOE was in effect.[9]

It was increasingly evident that the embargo was undermined by ubiquitous parallel market activities that the government was unable to curb, essentially for three reasons. First, army, police and customs officers were in no position to gain even a semblance of control over the vast Ethio–Somali border area, which stretched over more than 1,600 km. It is difficult for any government to maintain an effective control over such a vast stretch of land, but the Ethiopian government was particularly fettered. Contemporary correspondence reveals that the Customs Department had a limited number of poorly-paid officers and only a couple of aged vehicles, which frequently broke down.[10]

More importantly, official efforts were severely hampered by the fact that law enforcement officials were not able to single out and apprehend smugglers. For the Somalis, the political boundary was not a barrier to passage. As pointed out, Issa and Gadaboursi Somalis had long participated in exchange between Harerge and the coastal ports as intermediaries, providing transportation for merchants. In the process, they had acquired indispensable knowledge of geography and caravan routes and had established intricate, kinship-based commercial networks inside Somalia.[11]

9 HPA-GC.2, *Fit.* Demissie Tefera to *Dej.* Kifle Irgetu, Minister of State for Public Security, January 29, 1965. For theoretical consideration of the relations between increased risk and prices, see Christine Jones and Michael Roemer, 'Editors' Introduction: Modeling and Measuring Parallel Markets in Developing Countries,' *World Development*, 17:12 (1989), pp. 1861–70.

10 HPA-GC.2, Minutes of the Khat Commission of Inquiry, May 26, 1969; HPA-GC.2, Beshir Shiek Abdi, Governor of Jijjiga, to Harerge Provincial Administration Office, April 15, 1965.

11 Interviews: Yasin Usman, Dire Dawa, February 2, 1994; Mesfin Asfaw, Harer, March 15, 1994.

The capacity of authorities was further undermined by corruption, which allowed traders to engage in contraband activities with impunity. Law enforcement officials allowed smugglers to violate the terms of the SOE in exchange for graft. Corruption led to the proliferation of untaxed transactions and resulted in a sharp decline in government revenue, thereby diminishing the capacity of state institutions effectively to deal with tax evasion. The khat trade was not deemed a violation of the law before the restrictions were implemented. Even after SOE sanctions were imposed, transactions that took place west of Jijjiga were perfectly legal, and a few traders were permitted daily to bring in 200 kg of khat into the town. Khat transactions were illegal only east of Jijjiga. Consumers who resided east of Jijjiga were supplied by traders who took the risk of breaking the law. This is a clear case of a government defining the legality or illegality of commercial transactions.

Even though contraband trade had undercut the SOE, the official attitude toward the embargo did not change until November 1964 when Brigadier General Abebe Gemeda, Chief Administrator of the SOE, toured the areas under his command. While visiting Jijjiga the general observed that the embargo had succeeded only in undermining 'free trade,' thus denying farmers and traders the right to sell their khat in the open market and preventing the government from levying taxes on legal transactions. He noted that the embargo had not weakened Somalia's economy, thwarted subversive activities or incited rebellion in Somalia. Moreover, it was clear that the army was unable to secure the entire border and prevent smuggling, nor were intelligence officers able to weed out persons with subversive intentions from a population that was linguistically and ethnically akin to citizens of the purported enemy, Somalia. Put simply, the historical and contemporary realities of the region militated against the SOE's ability to achieve the expected political objectives.[12]

After observing the failure of the embargo, General Abebe made recommendations that marked a strategic shift in policy toward Somalia. The general argued for allowing an unrestricted flow of khat into Somalia. In his view, khat could be used best as a 'weapon' against Somalia's population and economy, not by denying consumers the substance they craved for, but by supplying it in unlimited quantities. He reasoned that khat was a harmful substance that would turn Somalia into a country of physically unfit and psychologically incompetent addicts who would not be able to field an effective fighting force. In addition, he reckoned that the cash Somalis spent on khat would devastate their country's economy by draining money into Ethiopia, which would also benefit from export taxes on the business. Based on these considerations, he recommended

[12] HPA-GC.2, Brigadier General Abebe Gemeda, Commander of the Third Infantry Division and Chief Administrator of the SOE, to Lt Colonel Tamirat Yigezu, *Inderassie* of Harerge, December 23, 1964.

that the khat embargo be lifted completely and immediately.[13]

In February 1965 General Kebede Gebre, the newly appointed *Inderassie* (the emperor's representative), concurred with some of General Abebe's analysis and abolished the 200 kg limit on Jijjiga's quota and permitted unlimited quantities of khat to be taken to the town. However, he left in place the ban on shipping khat to regions east of Jijjiga. In leaving the geographic restriction intact, the *Inderassie* argued that the free movement of people between the two countries would increase subversive activities against Ethiopia. The partial lifting of the ban, however, proved inconsequential, since most of the khat-consuming population resided in areas east of Jijjiga in such border towns as Togochale, Tafari Ber, Harshin, Dehatahil and Hartisheik, as well as in such Ogaden towns as Dagahbur, Kebri Dehar, Aware, Misraq Gashamo and Gode.[14]

Dissatisfied with the partial lifting of the embargo, General Abebe recommended that traders be issued municipal identity cards and be permitted to exchange an unlimited amount of khat within Ethiopia's borders.[15] Viewed against the backdrop of the embargo's negative economic impact on farmers, traders and the government, and its failure to achieve the political goals for which it was instituted, the proponents of the SOE were left with no justification for maintaining the status quo. In June 1965 a decision was made to extend the area in which khat was to be traded beyond Jijjiga, right up to the border with Somalia. The embargo against Somalia, however, was left in place as long as the SOE remained in force.[16]

Between June 1965 and October 1967 the khat business proceeded more or less smoothly. Officials were determined that the government should derive as much revenue as possible from the business. Although in principle Addis Ababa refrained from direct involvement in commodity production and marketing, the failure of the quasi-independent company that used to export khat to Aden left behind an enduring bias toward market-based exchange operations. Moreover, khat played too important a role in the finances of the state to be left free of government involvement. State intervention in the khat business needs to be understood in two ways: the level of revenue it extracted, and the means by

[13] HPA-GC.2, Brigadier General Abebe Gemeda, Commander of the Third Infantry Division and Chief Administrator of the SOE, to Lt Colonel Tamirat Yigezu, *Inderassie* of Harerge, December 13, 1964.

[14] HPA-GC.2, Lt General Kebede Gebre, *Inderassie* of Harerge, to Brigadier General Abebe Gemeda, Commander of the Third Infantry Division and Chief Administrator of the SOE, and *Fitaurari* Tewahade Woldeyes, Governor of the Harer sub-province and mayor of the city, March 7, 1965.

[15] HPA-GC.2, Brigadier General Abebe Gemeda, Commander of the Third Infantry Division and Chief Administrator of the SOE, to Lt General Kebede Gebre, *Inderassie* of Harerge, May 16, 1965.

[16] HPA-GC.2, Lt General Kebede Gebre, *Inderassie* of Harerge, to *Fitaurari* Tewahade Woldeyes, Governor of the Harer sub-province and mayor of the city, March 7, 1965; HPA-GC.2, Brigadier General Abebe Gemeda, Commander of the Third Infantry Division and Chief Administrator of the SOE to Lt General Kebede Gebre, *Inderassie* of Harerge, June 17, 1964.

which it did so. In the case of khat export from Harerge, two government departments, the Customs and the Internal Revenue, were charged with responsibility for collecting export duties and income taxes on domestic khat transactions.

Taxation, traders & the state

Before trade was disrupted by war in 1964, there existed two types of taxes on the khat trade. Export taxes were levied at a rate of 1.00 birr per kilogram of khat, while traders paid 0.50 birr in transaction taxes on domestic sales. Both rates applied to the entire Harerge province, whether khat was exported from Dire Dawa to Aden and Djibouti or through Jijjiga to Hargeisa in northern Somalia. During the period of the SOE, the collection of these taxes was suspended in the entire area east of Jijjiga. Khat export was permitted only in the Dire Dawa section, where the Customs Department continued to collect duties.

As we have seen earlier, traders were permitted to resume their domestic khat trade in June 1965, even though export to Somalia remained illegal under the SOE. Traders began to supply khat to such border towns as Tafari Ber, Togochale and Harshin, paying the 0.50 birr in taxes per kilogram of khat. The proximity of these towns to Somalia held out the possibility that khat could be smuggled across the border without payment of export taxes. For this reason, an additional 0.06 birr per kilogram export duty was imposed on all khat meant for consumption in regions east of Jijjiga. Since officially khat could not be exported to Somalia, the Customs Department had no role in the Jijjiga sector, and the nominal export duties (the 0.06 birr) were to be collected by the Internal Revenue Department. Traders were thus required to pay a total of 0.56 birr per kilogram in both income and export taxes.[17]

The export-duty rate reduction was meant to encourage legal trade and weaken the smuggling that had flourished during the period of the embargo.[18] The reduction of taxes by 0.44 birr was so attractive that many traders willingly paid the stipulated amount and operated legally. In the Dire Dawa section, northwest of Harer, the rate of both income and export taxes remained unchanged. The Customs Department continued to levy 1.00 birr per kilogram on khat that was exported to

[17] HPA-GC.2, Minutes of a Meeting of the committee assigned to study income tax conditions in Harerge, September 29, 1969.

[18] HPA-GC.2, Abate Astatke, Director of Jijjiga Customs, to Brigadier General Abebe Gemeda, Commander of the Third Infantry Division and Chief Administrator of the SOE, October 15, 1964.

[19] Before 1964 the combined rate of export and income taxes was 1.00 birr per kg. After the export tax rate was reduced to 0.06 birr, traders paid only 0.56 birr per kg of khat. The reduction amounted to 44%.

[20] HPA-GC.2, *Bejirond* Zewde Awoke, Treasurer of Harerge, to *Qegnazmatch* Tesfaye Wondimagegnehu, Governor of Gursum sub-province, January 26, 1966

Djibouti while the Internal Revenue Department collected income taxes at a rate of 1.00 birr per kilogram on domestic transactions.[19]

The new tax arrangement presented two major problems. First, legality ended once khat was cleared through the tax post at Harer, since most of the khat meant for domestic consumption was actually taken across the border after only 0.56 birr was paid on each kilogram. Second, for traders originating in areas east of Harer, the new tax structure allowed them a free ride to any destination. Traders who set out from Gursum *awraja* paid no taxes, whether they sold their khat within Ethiopia or in Somalia.[20] Smuggling was clearly depriving the state of a substantial portion of its rightful revenue. Customs officials estimated that the new policy resulted in a loss of 93,000 birr in just two months and argued for reinstatement of formal export taxes. But, as long as the SOE remained in effect, it was practically impossible to collect export taxes on the Jijjiga side.[21]

In January 1966 the provincial administration introduced two measures to prevent further revenue loss. The tax-collection post was moved from Harer to Jijjiga because the latter was deemed strategically situated to control the movement of traders. Gursum officials were authorized to levy taxes on khat that originated in their *awraja*. Income taxes were collected at Fugnanbirra, the capital of Gursum, at a rate of 1.00 birr per kilogram of khat. Under the new arrangement, traders bound for the Ogaden or northern Somalia were occasionally forced to pay income tax twice, once at Fugnanbirra and then at Jijjiga, before they reached their destinations.[22]

In October 1966 the provincial administration stepped in to thwart smuggling. A decision was made to equalize the income-tax rate paid at Jijjiga with that of the Dire Dawa sector, where income and export taxes had been 1.00 birr per kilogram of khat since 1960. The proponents of uniform taxation argued that contraband was nonexistent before 1964 when tax rates were the same throughout the province and called for the status quo ante in order to end unofficial trading activities. The nominal 0.06 birr export tax and the 0.50 birr income tax were consequently abolished and replaced by a flat 1.00 per kilogram of income tax regardless of the destination of the merchandise. In effect, taxes went from a total of 0.56 to 1.00 birr, whether khat was meant for domestic consumption or illegally exported to Somalia.[23]

The new rate was not popular with traders since, in determining the rate at which taxes were levied, officials apparently paid no attention to the fact that khat prices were seasonal and extremely volatile. One

[21] HPA-GC.2, Abate Astatke, Director of Jijjiga Customs to Brigadier General Abebe Gemeda, Commander of the Third Infantry Division and Chief Administrator of the SOE, October 15, 1964.

[22] HPA-GC.2, Lt General Kebede Gebre, *Inderassie* of Harerge to the Treasurer (*Bejirond*) of the Province, January 28, 1966.

[23] HPA-GC.2, A memorandum for Lt General Kebede Gebre, *Inderassie* of Harerge, October 24, 1966.

kilogram of khat, which sold, for example, for 5 birr during dry seasons, when only irrigated khat was available, would hardly fetch 2 birr during rainy seasons. Traders frequently complained about a tax regime that did not take into consideration the nature of the business, and they petitioned the government for a return of the tax rates to the 0.56 birr level. They also complained about the redundancy, confusion and inconvenience created by the decision to collect taxes in several places. Notwithstanding the problems, the khat business prospered and revenue collected at Jijjiga and Gursum rose sharply, reaching a comparatively high level of 1.6 million birr in 1967–8.[24]

Meanwhile, relations between Ethiopia and Somalia improved and occasioned the dissolution of the SOE. In a communiqué issued on September 18, 1968, the two countries declared that they had agreed to restore trade. The agreement provided for resumption of legal khat exportation to Somalia and meant that the Customs Department could begin to levy taxes at the old rate of 1.00 birr per kilogram. Accordingly, income tax was suspended in the Jijjiga section, since a double tax might have forced legal traders to go underground and lead to revenue decline. The previous system, of collection of income tax by the Internal Revenue Department and export duties by the Customs Department, was temporarily restored, and a commission was appointed to propose a uniform tax structure for the entire province.[25]

The commission thoroughly examined khat marketing, the relevant tax laws and practices of tax collection in Harerge. Based on its interpretation of the laws and its findings about marketing systems, transportation costs and price differentials between specific areas, the commission recommended adoption of a region-specific, rather than a province-wide, tax structure. Accordingly, in both Dire Dawa and Jijjiga sections, the export tax was fixed at 1.00 birr per kilogram of khat plus a 2 percent transaction fee. In the Jijjiga sector, so as to discourage smuggling, an additional income tax of 0.45 birr per kilogram was introduced.

In the Dire Dawa section, there were traders who used chartered planes to supply consumers in Ogaden towns such as Dagahbur, Kebri Dehar, Kelafo and Gode. The khat trade here was, therefore, different from other domestic operations, which used only vehicles. Taking into account that air transportation was more expensive and that khat prices were consequently higher in the Ogaden than in other places, traders who supplied this region were made subject to 0.55 birr per kg of khat, 0.05 cents higher than the income-tax rate stipulated for other domestic operations (Table 6.1).[26]

[24] HPA-GC.2, *Bejirond* Negatu Mekuria, Treasurer of Harerge, to *Dej.* Workineh Wolde Amanuel, *Inderassie* of Harerge, September 15, 1969.

[25] HPA-GC.2, *Dej.* Workineh Wolde Amanuel to *Fitaurari* Kebede Wadaju, May 13, 1970.

[26] HPA-GC.2, Lt General Kebede Gebre, *Inderassie* of Harerge, to the sub-provinces of Jijjiga and Degahpur, to the Director of Customs of the Northern Provinces and to Internal Revenue Administration offices of the sub-provinces, October 26, 1969.

Table 6.1 *Khat income tax & export duty rates as of October 28, 1969*

Sector		Customs (birr/kg)	Internal Revenue (birr/kg)	Transaction fee
Jijjiga	(export)	1.00	0.45	2%
	(domestic)	–	0.50	–
Dire Dawa	(export)	1.00	–	2%
	(domestic)	–	0.55	–

Source: Harerge Provincial Adminstration: General Corespondence, File 2 (HPA-GC.2).

Under the new tax structure, traders who exported khat to northern Somalia were required to pay a total of 1.45 birr in export and income taxes plus 2 percent transaction fee per kilogram of khat, or 158 percent more than the 0.56 birr rate. The initial reaction of traders was to submit petitions to the provincial administration, requesting that the new export duty and income tax rates revert at least to the rate in 1964, of 1.00 birr per kilogram. In their petitions, they put the provincial administration on notice that a strike would occur if their demands were not met.[27]

The end of the embargo was expected to lead to a sharp increase in the revenue collected by the Customs and Internal Revenue Departments. To the surprise of officials, collections declined, suggesting a major slump in the khat export business. By the end of November 1969, one month after the new tax rates were put into effect, the Customs and Internal Revenue Departments had collected a little more than 114, 000 birr, showing a 5.2 percent decrease from the 120,000 birr collected by the Customs Department alone during the same month of the previous year.[28] Regardless of the loss, the government pressed ahead with enforcement of the new tax rate. Neither the revenue drop nor the mounting petitions from traders were sufficient to nudge the government into revising its decision.

Toward the end of 1969 the official khat industry began to exhibit signs of dramatic decline. In the six months after the new taxes were implemented, the amount of export tax revenue collected by the Customs Department dropped from 830,053 birr to 542,512 birr, a decline of nearly 35 percent compared to the same months of the previous year. The loss in export taxes was somewhat compensated for by the reinstituted income-tax revenue, which brought in 239,202 birr (Table 6.2). The monthly breakdown of revenue statistics showed that

[27] HPA-GC.2, Appeal of 32 khat traders to *Dej*. Workineh Wolde Amanuel, *Inderassie* of Harerge, November 10, 1969. Twenty of the 32 petitioners were women. HPA-GC.2, Appeals of 9 khat traders to the Office of the Governor of Jijjiga sub-province, May 3, 1970.

[28] HPA-GC.2, Beshir Shiek Abdi, Governor of Jijjiga, to Colonel Yilma Teshome, Commander of the 10th Infantry Division and Chief Administrator of the SOE in the Northern Section of the Province, April 4, 1970.

the loss in tax revenue increased each month, clearly signaling that the khat industry was facing a serious problem.[29] The fact that income taxes registered a consistent increase while export duties spiraled downward month after month unambiguously indicated that more traders were opting for the lower income-tax rates, although it cannot be said that they were not smuggling.

Table 6.2 *Comparative statistics of export duty (A & B) & income-tax revenues in birr*

Months	1968–9(A)	1969–70 (B)	Difference (A & B)	Income taxes
Oct.18–30	63,832	37,333	26,499	16,462
November	120,100	79,287	40,812	34,953
December	127,982	91,801	36,180	40,474
January	156,391	108,524	47,866	47,852
February	176,770	117,862	58,908	51,972
March	184,975	107,701	77,274	47,488
Total	830,053	542,512	287,541	239,202

Source: HPA-GC.2, *Bejirond* Belete Gebre to Harerge Provincial Administration, April 19, 1970.

Quantitative evidence is not sufficient to identify the specific reason for the drain on government revenue, but there is documentation indicating that the number of licensed export traders was consistently declining, suggesting a weakening official export market. In October 1969, just before the new tax rates were implemented, there were more than 100 licensed private traders in the khat export business. By April 1970 there were only five such traders, all of whom were women.[30] In May all khat traders, most of whom were Somalis, declared that they had stopped exporting khat to Somalia, claiming bankruptcy and citing the government's failure to act on their petitions to repeal the new tax rates.[31] Authorities interpreted the action of the traders as a 'strike'[32] and even suspected sabotage by Somali government agents. The khat issue had obviously become a political problem.

In fact, the authorities were never convinced that the traders' petitions were genuine appeals for reductions of unaffordable tax rates. Officials

[29] HPA-GC.2, Belete Gebre, Director of Customs of the Eastern Regions, to the Office of Harerge's Provincial Administration, April 19, 1970.

[30] Women submitted the majority of petitions to the provincial administration requesting the resumption of the 'Hargeisa' route. Women retailed khat in local markets, traveling only short distances from the town of Jijjiga. The prohibition deprived women of their independent source of income and forced them to revert to their traditional role of caring for the family. Men were not quite as interested in the resumption of trade since they were able to engage in contraband trade while women were obligated to stay close to their families.

[31] HPA-GC.2, telegram from Assefa Aboye, Director of Jijjiga Customs, to *Dej.* Workineh Wolde Amanuel, May 13, 1970.

[32] The Amharic term for a strike, *adma*, connotes conspiracy, not just a strike in demand of small adjustments in wages, working conditions and so forth.

were persuaded that khat traders were 'getting rich' by defrauding the government, primarily because of smuggling that was allowed to go unchecked. The traders' conspicuous display of wealth, such as vehicle purchases for private and business use, only confirmed their suspicions and led to their insistence on maintaining the new tax rates. Revising the tax rate was never considered as a means to curb contraband trade activities. Most officials instead advocated cracking down on the smugglers.[33]

Even though the government was committed to increasing revenue and stamping out parallel market exchanges, it did not possess the institutional, logistical and financial capacity necessary to secure the border areas and control the activities of contraband traders. Underpaid, poorly trained and inappropriately equipped government personnel were unable to curtail the activities of experienced and well-organized contraband traders. The latter's brand new, faster and dependable pick-up trucks were able to traverse the sandy, flat lands between Jijjiga and the Somali border with little or no interference from the government. Official documents from subordinate officials are full of complaints about the lack of personnel, finance, communication equipment and dependable means of transportation in the fight against illegal trade. In April 1970, just before khat traders went on a 'strike,' there were only twelve police officers in Jijjiga, equipped with only one poorly maintained vehicle.[34]

It is probable that the new tax rate was so high that it forced licensed traders to opt for parallel markets. In the early 1970s smuggling obviously owed a great deal to the traders' increased capacity to evade customs posts. Licensed traders also shrewdly manipulated the new tax structure to expand the parallel market for khat. The new tax-collecting arrangement gave overlapping jurisdiction to the Customs and Internal Revenue Departments to levy their respective taxes at Jijjiga. The arrangement was meant to reduce the chances of traders' crossing over to Somalia without paying export duties, but apparently no amount of tinkering with the taxation structure had much effect. Traders were still able to evade taxes and smuggle khat into Somalia (Map 6.1).

After purchasing khat in the vicinity of Harer, traders headed directly for the customs post at Jijjiga, labeling their consignment 'for domestic consumption.' Several hundred kilograms of khat daily passed through the tax-collection points after income taxes were paid at the rate of 0.50 birr per kilogram as required by the new law. Past Jijjiga, it was almost impossible to track down and ensure that the leaves were consumed within the limits of the declared destination. In fact, most of the khat that was ostensibly intended for domestic consumption was transported to Somalia by three routes. A few miles out of Jijjiga, the first route

[33] HPA-GC.2, Report of the Khat Commission of Inquiry, submitted to *Dej.* Workineh Wolde Amanuel, May 16, 1970, p. 6.

[34] HPA-GC.2, Beshir Shiek Abdi, Governor of Jijjiga, to Colonel Yilma Teshome, Commander of the 10th Infantry Division and Chief Administrator of the Northern Section of the SOE, April 4, 1970.

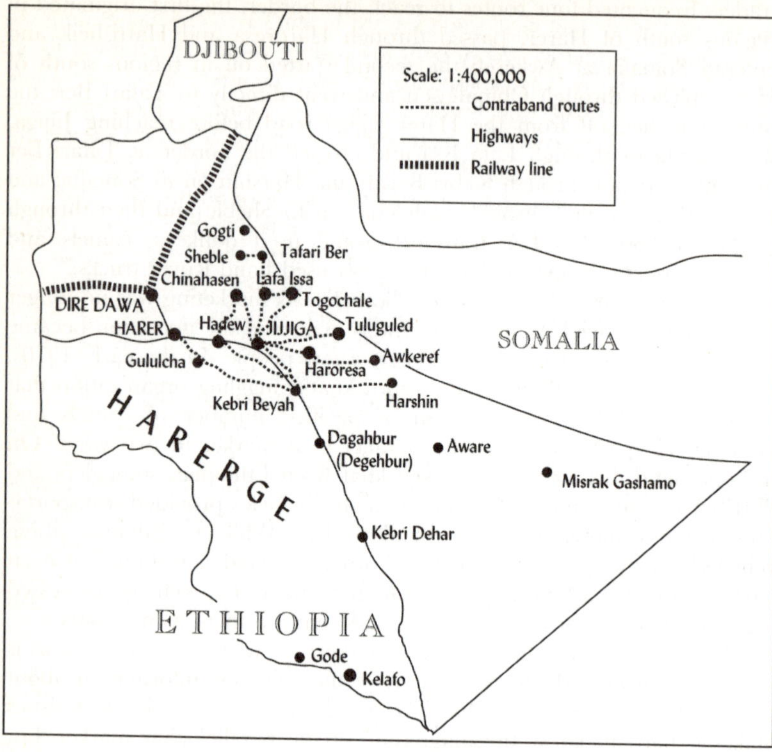

Map 6.1 *Khat smuggling routes,* 1960s

branched off south and went to Kebri Beyah, where it broke into several roads leading into Somalia. The second turned northward to Lafa Issa, where it diverged in different directions and crossed the border. Still others remained on the Jijjiga–Hargeisa route, proceeding to Togochale and crossing into Somalia through the guarded border points. In such cases, officials allowed the traders to transfer khat to their clients from Somalia in return for graft.[35]

Unlicensed traders participated in the parallel markets but pursued different strategies to 'beat the system.' The unlicensed traders were not allowed to operate officially, that is, they were not permitted to clear through the tax-collection post at Jijjiga. Thus, they avoided Jijjiga and the tax-collection post there altogether and reached the border crossing points without paying taxes of any kind. They therefore differed from the licensed traders, who usually paid the lower rate of income tax, although in most cases they smuggled the khat out. The unlicensed

[35] HPA-GC.2, Report of the Khat Commission of Inquiry, submitted to *Dej.* Workineh Wolde Amanuel, May 16, 1970, p. 6; HPA-GC.1, Beshir Shiek Abdi, Governor of Jijjiga, to Hareroe Provincial Administration Office, June 7, 1970.

traders frequented four routes to reach the border: the first originated in regions south of Harer, passed through Haroresa and Hartisheik and entered Somalia at Awkeref; the second started off in regions south of Harer, passed through Chinahasen and went directly to Tafari Ber; the third branched off from the Harer–Jijjiga road before reaching Jijjiga, went northeast through Lafa Issa and crossed the border at Tafari Ber or went southeast through Kebri Beyah and Harshin on to Somalia; and the fourth proceeded through Gululcha, on to Sheble and then through Gogti into Somalia. For transport, some used donkeys, camels and human porters to transport khat, but most used Land Rover trucks.[36]

Once the commodity was inside Somalia, its marketing was no longer considered unofficial. Import duties were paid and the operation became a well-organized process controlled by a few people. In the early 1970s, a certain Abdi Sheriff reportedly ran a khat-smuggling organization that owned 28 Land Rovers and an unspecified number of camels and donkeys. He participated in the parallel khat market in two ways. On the one hand, his agents purchased khat from Ethiopian smugglers and distributed it in Somalia. On the other, his vehicles provided transportation within Ethiopia for contraband traders. While in Ethiopia, these vehicles usually operated at night in order to avoid detection, but even daylight was not a hindrance to their movement. The vehicles possessed registration plates for both countries and changed them as necessary.[37]

The infrastructure of the parallel market was well organized but, as is the case with all hidden exchanges, we have far less information about the unofficial transactions than we would like. The lack of statistics makes it hard to know the magnitude of the parallel khat market, but estimates may be drawn from the figures of the official market. In the late 1960s, 7,000–8,000 kg of khat were officially exported to Somalia every day.[38] Circumstantial evidence suggests that the quantity of khat exported through unofficial channels was greater. For instance, the amount that daily cleared through Jijjiga as a supply for residents of Ogaden towns reportedly exceeded their need by more than 5,000 kg.[39] This figure excludes the khat that never reached the customs post at Jijjiga, having been taken by smugglers directly to the border crossing points. If we assume that the quantity of the smuggled khat was half the official amount, a combined total of approximately 12,000 kg (8,000 official and 4,000 unofficial) of khat entered Somalia daily. Following the

[36] HPA-GC.2, Report of the Khat Commission of Inquiry, submitted to *Dej.* Workineh Wolde Amanuel, *Inderassie* of Hararge, May 16, 1970, pp. 6–7.

[37] HPA-GC.2, Belayneh Abebe, Acting Director of Jijjiga Customs, to *Dej.* Werkineh Wolde Amanuel, *Inderassie* of Hararge, May 10, 1970; HPA-GC.2, *Dej.* Workineh Wolde Amanuel to Major General Abebe Gemeda, Commander of the Third Infantry Division, May 12, 1970.

[38] HPA-GC.2, Telegram from the Governor of Jijjiga to Col. Yilma Teshome, Major General Abebe Gemeda, *Dej.* Workineh Wolde Amanuel and *Bejirond* Negatu Awoke, October 18, 1969.

[39] HPA-GC.2, *Fit.* Meharene Minda, Deputy *Inderassie* of Harer, to the Third Infantry Division Command, October 19, 1969.

'strike' of May 1970, no khat officially cleared through Jijjiga and consumers in Somalia were in all likelihood supplied through unofficial channels. It is thus possible to estimate that approximately 360 tons of khat were smuggled into Somalia monthly during the 'strike.'

In Somalia, a kilogram of khat fetched as much as 5.00 birr during low supply seasons. At an average price of 2.00 birr per kilogram, Somali consumers spent an estimated 9 million birr on khat annually. The high value suggests that, by the early 1970s, the khat industry had become the main source of income and employment for farmers, traders and those who provided auxiliary services as truck drivers, porters, packers and brokers.[40] As will be shown in Chapter 8, hundreds of people also depended on the proceeds from khat to purchase consumer commodities in Somalia to smuggle into Ethiopia. The khat business was so central to Hararge's economy that a disruption would inevitably weaken the official market, strengthen parallel ones and adversely affect the region's political and economic well-being.[41]

Fear of dislocation and political instability made the khat issue a matter of national concern. An emergency meeting attended by *Ato* Mulatu Debebe, the Minister of State for Finance, Colonel Solomon Kedir, the Vice Minister of State for Customs and *Dejazmatch* Workineh Wolde Amanuel, Hararge's *Inderassie*, planned and launched an effort to encourage traders to resume their activities and avert political unrest. The plan approached the problem of smuggling from two important, if contradictory, standpoints. The first grew out of the thinking that a more organized, well-equipped and better-financed police force would gain control over the contraband trade. The second was born of a reluctant recognition of the need to reform the high tax regime, which had become onerous, redundant and cumbersome. In regard to the first approach, a campaign against contraband trade was immediately launched, while the issue of taxes was referred to a commission set up to study the problem.[42] To implement the first part of the plan, which emphasized physical controls, customs officers, Ministry of Finance police and army troops were brought in to Jijjiga from Harer Zuria, Gursum and Wobberra *awrajas* and were deployed along the major smuggling routes, specifically in the towns of Togochale, Lafa Issa, Haroresa, Mekanisa and in the Dahital–Hartisheik area. Six transport vehicles were assigned to the service of the task force.[43]

The bulk of official correspondence on the subject reveals that the campaign was ineffectual from the outset. The Ethiopian government

[40] Demissie, *The Kottu of Hararge*, p. 92.

[41] HPA-GC.2, Report of the Khat Commission of Inquiry, submitted to *Dej.* Workineh Wolde Amanuel, May 16, 1970, p. 1.

[42] HPA-GC.2, *Dej.* Workineh Wolde Amanuel to Mammo Taddesse, Minister of Finance, May 10, 1962, E.C; HPA-GC.2, Lt Colonel Solomon Kadir, Vice Minister of State for Customs, to Mammo Taddesse, Minister of Finance, May 18, 1970.

[43] HPA-GC.2, Minutes of the meeting of the committee to control contraband trade, Harer, May 13, 1970; HPA-GC.2, Belayneh Abebe, Acting Director of Jijjiga Customs, to *Dej.* Workineh Wolde Amanuel, May 10, 1970.

did not have the resources to overwhelm the contraband traders. Letters written at various times during the three years of the campaign illustrate the point. One written immediately before the traders' 'strike' of May 1970 states:

> The magnitude of khat transiting through contraband channels is considerable. It has weakened the legal khat market and undercut those traders who [abide] by the law, preventing them from fulfilling their legal obligations and operating officially. It is important that the vehicles that were assigned to the Ministry of Finance police be sent [to us] as soon as possible. [If this is not done] the 'normal' traders would join the contraband sector and government revenue would disappear.[44]

Another letter, written two months into the 'strike' and a month after the start of the campaign, states:

> Members of the Ministry of Finance police who were assigned to control the khat wealth that is being smuggled out ... have no vehicle effectively to patrol the area and they have not been paid per diem [since the campaign got underway]. Contraband traders have taken advantage of the lapse and have increased their activities. [Please] dispatch the vehicles that were reportedly purchased for the purpose [of the effort to stop contraband trade] and make the per diem paid to the police personnel as soon as possible.[45]

A third letter, written three years after the 'strike' and the start of the campaign, by an apparently frustrated police commander, reminded high officials that it was 'impossible for a police officer who has only two feet to chase and catch a khat-loaded, "four-footed" vehicle which has escaped without paying taxes.'[46] Clearly, the policy of physical controls was not effective because the smugglers 'beat the system,' which did not possess the efficiency, equipment and personnel necessary to gain control over the border. The first approach, therefore, failed to force traders into the official sector.

Meanwhile, the Khat Commission of Inquiry began its investigations in May 1970 to determine the causes of smuggling and of the traders' 'strike.' The commissioners toured the porous border areas, carried out extensive interviews with elders, farmers, traders (mostly women) and government officials. The commission's final report cited higher taxes, the loopholes in the newly implemented tax-collection system and the contraband traders' improved transport capability as the principal reasons for the growth of smuggling and the consequent decline in tax revenues.[47]

[44] HPA-GC.2, *Fitaurari* Beshir Shiek Abdi to Col. Yilma Teshome, *Dej.* Workineh Wolde Amanuel, Major General Abebe Gemeda and *Bejirond* Negatu Makuria, April 19, 1970.

[45] HPA-GC.2, Colonel Yilma Teshome to Colonel Solomon Kedir, *Dej.* Workineh Wolde Amanuel, Major General Abebe Gemeda, *Bejirond* Negatu Makuria, the Director of Customs for Eastern Provinces and the Governor of Jijjiga, July 19, 1970.

[46] HPA-GC.1, Commandant of Ministry of Finance police of Wobberra sub-province to Office of Wobbera's Treasurer, February 2, 1971.

[47] HPA-GC.2, Report of the Khat Commission of Inquiry, submitted to *Dej.* Workineh Wolde Amanuel, *Inderassie* of Harerge, May 16, 1970, pp. 1–10.

The views of traders, farmers, local leaders and government officials, which were quoted verbatim in the minutes of the commission's deliberations, bear out the fact that higher tax rates precipitated the traders' 'strike.' Furthermore, a work sheet showing the expenses borne by traders as khat traveled from producer to consumer, also included in the commission's report, revealed that the new tax rates absorbed all potential profits. Even though the evidence showed that tax rates were responsible for the 'strike,' the commissioners concluded that it was the unchecked expansion of contraband trade that had frustrated traders and forced them out of business. They recommended, predictably, bolstering physical controls and improving the government's institutional and logistical capabilities to secure the border and apprehend and punish illegal traders.[48]

The then Minister of Finance, Mammo Taddese, rationalized that cunning traders had brilliantly manipulated the rate differential between income taxes and export duties to evade taxes, thereby increasing the quantity of khat that was smuggled into Somalia. At Jijjiga, traders paid the lower income tax and cleared the tax-collection post, pretending to be suppliers for domestic consumers, but mostly transported their khat into Somalia. Since they avoided paying the higher export duty, the Finance Minister reasoned that the logical solution to the parallel market problem was to make the abused income-tax rate prohibitive and reduce export duties in order to encourage legally exported khat. Accordingly, he reduced the export tax from 1.45 birr to 1.25 birr and raised income-tax rates from 0.50 birr to 0.65 birr per kilogram of khat.[49]

Because the net change was insignificant (0.05 birr), these adjustments did not change the behavior of traders. The effort to secure the border, as pointed out earlier, was ineffectual, and traders preferred the still lower income taxes (at 0.65 birr) to the export duties (1.25 birr). Careful examination of the records of the provincial administration from 1970 to 1974 reveals that the cat-and-mouse game between the government and traders, which was actually a competition over the khat industry, was as intense as ever. Interdepartmental correspondence shows the continued inability of government to secure the border and to curb illegal commercial activities owing to the inadequacy of equipment and personnel.

In terms of achieving control over parallel markets, the government did not match the resources of contraband traders. Nor did the negligible export-tax reduction convince khat exporters and domestic dealers to return to the official sector. When its effort failed in the border areas, the government resorted to squeezing as much revenue as possible from small khat retailers who operated in urban centers. These were mostly women, who had never been required to pay taxes. Provincial archives from the early 1970s are replete with reports of intimidation, physical

[48] HPA-GC.2, Minutes of the deliberations of the Khat Commission of Inquiry, Jijjiga, May 14, 1970, pp. 1–53.

[49] HPA-GC.2, Almazneh Workie, Vice Treasurer of Harerge, to the Internal Revenue Departments of Harerge's seven sub-provinces, July 9, 1970.

harassment and imprisonment of small traders on account of their failure to pay taxes.[50]

In the early 1970s the khat business in the Jijjiga sector faced new and serious challenges from the Somali side. Following the October 1969 military coup d'etat in Somalia, demand for khat dropped, perhaps owing to a general economic slowdown caused by political uncertainties and the fact that unpaid civil servants were unable to buy khat. The main reason for declining demand was nevertheless the increase of import duties on khat from Somali shillings (Ssh) 6.06 to Ssh 10.50 per kilogram.[51]

The decision to raise import duties to prohibitive levels was part of the new Somali government's stance against the consumption of khat, particularly that which was imported from Ethiopia. The military leaders capitalized on the general popular dislike of Ethiopia to end a habit they considered harmful to Somali society and the economy. In November 1971 a cholera epidemic was reported in Ethiopia, and the Somali government seized the moment to suspend importation of khat. The ban was lifted two months later, but future imports were placed under strict government control. All the khat imported from Ethiopia was directed to one place, Nebodid, where government inspectors examined quality and customs officials levied duties at the rate indicated above.[52]

In the Jijjiga sector, khat was a pawn in the contentious political relationship between Ethiopia and Somalia. Depending on circumstances, each government had used khat as either a source of revenue or as a 'weapon' against the other's economy. Political considerations more often than economic merits influenced policies concerning the marketing of khat. In the early 1970s Harerge's khat economy suffered from a coincidence of policies in both Ethiopia and Somalia that centered on controlling the movement, marketing and use of the commodity. The evidence from the Ethiopian side shows that government intervention in the marketing of khat remained a problem until the mid-1970s. By the time of the imperial regime's downfall in 1974, despite repeated appeals from traders for reductions,[53] tax rates had remained high and parallel market activities had not been brought under control.

The difficulties of the Jijjiga market diverted traders to the Dire Dawa section. Since the late 1920s, when khat began to be exported in significant quantities, Dire Dawa had been the main center for purchase,

[50] HPA-KF.1, Women khat retailers in Harer town to the Provincial Administration Office, March 10, 1974; HPA-KF.1, A petition signed by 32 representatives of Harer women protesting higher taxes to the Provincial Administration Office, August 30, 1974; HPA-KF.1, Women khat traders in Dadar town to the Administrator of Wobberra *awraja*, May 7, 1974; HPA-KF.1, Women khat traders in Jijjiga town to the Administrator of Jijjiga *awraja*, September 28, 1975.

[51] HPA-GC.2, Belete Gebre, Director of Customs for Eastern Provinces, to *Dej.* Workineh Wolde Amanuel, November 14, 1970.

[52] HPA-GC.2, *Fit.* Meharene Minda, Deputy *Inderassie* of Harerge, to *Bitwoded* Zewde Gebre Hiwot, Minister of Interior, November 19, 1970.

[53] HPA-GC.2, Nine licensed khat traders to the office of Governor of Jijjiga sub-province, May 3, 1970.

processing and shipment to Djibouti. The market did not suffer the difficulties that had beleaguered Aden and Hargeisa, perhaps because of its efficiency. Dire Dawa's khat export was dominated by a few well-organized private exporters who also controlled the market in Djibouti. In the 1960s and 1970s the businesses were technically privately owned enterprises, but the owners operated through a Djibouti-based *Association des Importateurs et des Marchands de Kath* (AIMK). By 1975 the consortium had 51 members who maintained agents in Dire Dawa.[54]

The consortium managed to keep competitors from gaining ground in the Dire Dawa market, primarily owing to its exclusive access to air transportation and its distribution network in Djibouti. The AIMK contracted with Ethiopian Airlines daily to dispatch 3,600 kilograms of khat, and the airline agreed not to accept khat shipments from other exporters. Once khat arrived in Djibouti, members of the consortium received their share of khat and distributed it to retailers.[55]

In addition to their monopoly over transportation and the market in Djibouti, the AIMK also maintained unique business relations with khat farmers in Kombolcha and the northeastern part of Haramaya where the *daalota* khat variety, popular in Djibouti, grew abundantly. Agents bought khat on the farm, offered farmers prices higher than the market and furnished their own transport to Dire Dawa. The exporters thus established special relations with producers. With control of production, access to air transportation and distribution networks in Djibouti, the consortium established a near-monopoly position in the Dire Dawa sector.[56] In the mid-1970s a few traders unsuccessfully attempted to establish a competing organization in Dire Dawa.[57] They failed for three reasons: first, producers refused to sell khat to the new 'company's' agents;[58] secondly, the new exporters were unable to penetrate the market in Djibouti, since it was controlled by the AIMK; thirdly, the new 'company' could not get Ethiopian Airlines to transport its khat.

Supplied by air, Djiboutians consumed fresh, potent but expensive khat. High-price commodities often favor the rich, but, in Djibouti, even the poor were willing to spend a substantial amount of their income on khat. In the 1960s Djibouti annually spent 800 million DF on khat, or the equivalent of half the territory's revenue. In 1965 some youth clubs carried out a week-

[54] HPA-GC.2, Mohamed Sougal Bahdon, Président de l'Association Djibouti, to the Director General of the Ministry of Commerce, Dire Dawa, November 28, 1975.

[55] HPA-KF.1, Manager of Operations of Ethiopian Airlines to the branch office of the Ministry of Trade, Industry and Tourism in Harerge, April 10, 1974; HPA-KF.1, Khat Traders in Dire Dawa to the Minister of Trade, Industry and Tourism, October 22, 1974.

[56] HPA-GC.1, A petition from the Farmers of Dire Tiara, Kombolcha district, addressed to the Governor of Harer Zuria *awraja*, August 2, 1974.

[57] HPA-GC.1, The appeal of six traders to Samuel Teferra, Governor of Dire Dawa sub-district and mayor of the town, April 27, 1975.

[58] HPA-GC.1, Hajji Abdul-Rahiman Ali and Hajji Bakar Ali, representatives of the people of Kombolcha, to *Fitaurari* Kebede Wedaju, Vice *Inderassie* of Harerge, August 31, 1975.

long campaign, with the assistance of the mass media, against khat use, but their efforts did not diminish the habit. A year later Djibouti's Governor Saget forbade the importation and sale of khat in the territory and ended up provoking popular discontent.[59] Khat consumption did not show any sign of tapering off as the result of these actions.

Conclusion

We have seen how the Ethiopian government sought control over khat-generated wealth and the ways in which traders resisted policies they considered intolerable state intervention in agricultural marketing. But the production and marketing of khat was a big business that provided opportunities for profit; the growth in scale of the business provided an unparalleled occasion to enhance state revenues and prompted attempts to control it, first through trading monopolies and later through taxes and duties.

There were several instances of a concerted action aimed at improving the conditions of trade, including the unprecedented feat of successfully organizing a 'strike.' In the main, however, the challenge traders presented against unpopular policies centered on trying to circumvent or bypass the machinery put in place to extract revenues. The policy to impose and increase tax rates led to a new growth industry, smuggling. In their efforts to avoid what they thought to be unfair obligations to the government, traders were aided by institutions such as kinship and indigenous commercial networks that the government was not able to penetrate. Still, the traders' main advantage was the government's inability to implement its own policies.

Even though analysis of the capacities of the Ethiopian state or even of its policies on khat is beyond the scope of this study, the policies that were enacted and the ways the government went about implementing them raise important questions about state capacity. The struggle between the government and traders over khat marketing showed that the Ethiopian government did not have the resources necessary to implement its policies, especially in terms of controlling contraband. The series of policy measures that the government took became objects of a struggle, and groups not favored by the state did everything to avoid government control and the implementation of its policies.

It is apparent that policy formulation does not guarantee its effective implementation. Still, as Joel Migdal states, 'social control for the state entails more than insinuation or penetration of its agencies into society, more even than just successful extraction of resources. It includes the ability to appropriate resources for particular purposes and to regulate people's daily behavior.'[60] Given the fact that the imperial state was less

[59] Thompson and Adloff, *Djibouti and the Horn*, pp. 164–5.

[60] Joel S. Migdal, *Strong States and Weak States: State–Society Relations and State Capabilities in the Third World* (Princeton, 1988), p. 61.

than able to implement its khat policies, it may appropriately be labeled as 'weak.' The difficulties faced by the export market, ranging from market monopoly to government intervention, appear to have created tremendous inconveniences for traders but brought little change to the production levels of the commodity. In addition to unofficial markets, domestic consumption absorbed the surplus created by the demise of official markets, thus increasing its role in the overall khat market in the 1960s and early 1970s.

Seven

Domestic Use & Trade

Social mobility, structures of distribution & self-sufficiency 1964–74

> Everything in this town was built by the farmers who left their farms and moved here. They built houses, stores, restaurants, gas stations and so forth. AwWoday is a town of farmers who are still moving in and building their residences and businesses.
>
> – Alemayehu Woyessa, 2000

The decade preceding the Ethiopian revolution of 1974 witnessed a rapid expansion of khat consumption in Harerge. Since no systematic consumer survey has ever been undertaken, there is no quantitative data to document the process, but circumstantial evidence suggests that consumption greatly increased during the 1960s. Older informants generally perceive a dramatic contrast between the widespread consumption of khat during that decade and pre-Second World War levels. Middle-aged informants recollect the proliferation of khat parties in most towns in the region. By the early 1970s, more men and some women were spending afternoons chewing khat with friends in private homes and tea houses; it was common to see construction workers, students, truck drivers and cabbies, laborers and shopkeepers, irrespective of religious affiliation, chewing khat. In the 1940s and 1950s, this had been a much rarer sight.[1]

These recollections are supported by written evidence. In 1960 it was reported that 'Ethiopian Christians [in Harerge] abhor[red] the leaf, … . [In contrast, khat occupied] a position of considerable importance … among the Moslem agricultural Eastern Galla [Oromo].'[2] A decade later a different picture was painted: khat 'use has spread so much that, in Harerge Province, many age groups of both sexes, Christian as well as Muslims, consume[d] a lot of chat [khat].'[3] Confirming this observation

[1] Interviews: Girma Andarge, Dire Dawa, February 23 and 27, 1994; Duressa Feyissa, Dire Dawa, February 24, 1994.

[2] Brooke, 'Khat in the Middle East,' p. 50.

[3] Demissie, *The Kottu of Harerge*, p. 95.

it was reported in 1973 that [khat-chewing] 'now cuts through many faiths, social levels, and age groups... . Men and women on the farm and people in towns, whether they be civil servants or business men, are to be seen taking time to chew chat [khat]. The number of school boys chewing it is very high.'[4] In a matter of a decade, therefore, khat-chewing had become a ubiquitous practice involving all of Harerge's demographic divisions.

Until the late 1960s, most of the areas where consumption increased were under a State of Emergency and trade was prohibited. The demand created was met by smuggling. The market outlet, albeit illegal, made it possible for khat to remain part of the cropping system and an important cash earner for farmers. Nevertheless, smuggling was unlawful and the market opportunities it created were risky, unreliable and unsustainable. In response, farming families adopted new strategies, which included intensive use of resources and diversified cultivation of crops that could be used for both market and consumption purposes. Smuggling also led to the emergence of new internal marketing structures and necessitated modifications in production strategies.

Domestic consumption & proliferation of market centers

One marked indication of the increase in khat consumption in the 1960s was the expansion of khat markets in both size and number. The Dire Dawa retail market became second only to that of Harer, although the latter was frequented by retailers who sold their khat directly to urban consumers. Distribution was facilitated by the completion, in 1963, of the Miesso–Dhangago road, which was constructed to connect the Harer–Dire Dawa highway to the coffee-producing regions of Chercher. The widespread use of motor vehicles on these roads enabled producers in distant areas of Chercher to bring their khat to the Dire Dawa market.

The importance of the Miesso–Dhangago road was enhanced by the construction of two feeder or farm-to-market roads, which connected Asba Tafari to Gelemso and Kobo to Dadar. Jitneys, pick-up tracks and automobiles were able to make increased use of the road, enabling farmers to reach the markets in Dire Dawa and Harer faster and inexpensively. For instance, after the completion of the road, the cost for transporting a quintal of khat from Kobo to Dire Dawa dropped from 8.00 birr to 3.00 birr. The new roads enabled traders to travel to production areas and purchase khat in local markets and along the road. In the early 1970s farmers had told researchers that the road had

[4] Amare and Krikorian, 'Coffee's Rival,' p. 371.

[5] Lakew Birke, *A Study of the Development Impact of the Miesso-Dengego Road*, Department of Agricultural Economics and Business, College of Agriculture, Haile Sellassie I University (Alemaya, 1973), pp. ii–iii, 15, 30, and 64.

helped them only slightly to increase khat production.[5]

Without the road and faster transportation, khat production could not have increased at all. The completion of the road coincided with Aden's rejection of Ethiopian khat and the diminution of demand for the commodity. It could be said that the road moderated the effects of the loss because increased accessibility coupled with faster and convenient transportation augmented local consumption and compensated for the export market. Yet falling transportation costs did not cause a decrease in khat prices or much financial benefit for farmers because the market was dominated by a few merchants based in Dire Dawa who, as exclusive exporters to Djibouti, possessed the power to set prices.

During the second half of the 1960s new khat markets appeared throughout the rural areas and others were enlarged. In the years before 1964, officials of CHATHA (the company exporting to Aden), preferred to purchase khat in Dire Dawa in order to shift to farmers the cost of damage to khat caused by the speedy transport on bumpy roads. By staying in Dire Dawa, however, purchasers grew unable to compete for the best-quality khat with independent buyers who offered higher prices on farm sites. Eventually the company failed to meet the demands of Adeni consumers in terms of both the quality and quantity of khat needed there.

The Djibouti exporters made a conscious decision to avoid the mistakes of the defunct CHATHA by maintaining the quantity and quality of khat delivered each day. For them, sustaining the market was more important than trying to maximize their profit. The exporters dispatched buyers to the highlands to purchase khat directly from farmers. Haramaya, a town located midway between Dire Dawa and the producing regions, was then the largest wholesale khat-buying market in the Harerge highlands. Donkey-loads of khat were carried here from the surrounding areas, and, after a brisk morning market, khat was transported east to Dire Dawa and west through Harer to Jijjiga.[6]

Throughout the 1960s Haramaya remained the largest khat market. Meanwhile, at a place called AwWoday on the Harer–Dire Dawa road, an enterprising Somali started a small café to cater to travelers to and from Harer. Soon a small but bustling khat market sprang up next to it. Small grocery shops were set up so that farmers could return home with necessities bought with the proceeds of their khat sales. The café's site was quickly transformed into a commercial town, which, by the late 1960s, had become a large center where khat was purchased and distributed throughout the province.[7] Produce was brought there from Haramaya and Kombolcha districts and from Dadar and Wobberra or Gara Mula'ata areas. Most of the khat that originated in Haramaya–Kombolcha areas went to Dire Dawa and thence to Djibouti while leaves from Wobbera–Gara Mula'ata districts were dispatched to Jijjiga,

[6] Interview: Luye Said, Haramaya, Ganda Limay, May 26, 1994.

[7] Interview: Mumed Ahmed Umer, Haramaya, May 17, 1994.

whence the khat was officially and unofficially exported to northern Somalia.[8]

AwWoday quickly surpassed Haramaya as the leading khat market center. A study conducted in 1973 on the marketing of agricultural products in Haramaya district shows that 85 percent of the khat marketed that year was exchanged in AwWoday, while 10 and 15 percent were sold in large city markets and on the farm, respectively.[9] Informants suggest that, while Haramaya was preferred to traveling to Dire Dawa, it was not a place for a khat market to flourish. The fact that municipal officials taxed khat transactions within the city was cited as one reason that caused buyers and sellers to avoid the city. The main reason for the decline of Haramaya was the perception of the Muslim producers and marketers that the town had become incompatible with khat. As an administrative center, Haramaya was populated by Christian 'Amhara' civil servants, who preferred to consume *tella,* a local ale, rather than engaging in khat-chewing. Pubs were opened to cater to these residents, and people usually got drunk there. Informants insist that khat was too sacred a leaf to mix with alcoholic beverages. Thus, the Muslim khat farmers felt duty-bound to shift to the new marketplace at AwWoday, away from the exacting municipality and the 'defiling *tella.*'[10]

As the leading market center for khat, AwWoday became a destination for young men looking for new opportunities. Migration to urban centers became desirable because it was creative, adventurous and future-oriented, and it helped reduce demographic pressure on land in rural areas. By the 1960s population densities in rural areas were so high that farming had become less and less attractive. In response, siblings often made arrangements by which some of them migrated to towns to retail khat, a percentage of whose proceeds was remitted to those who remained on the farm. Khat-selling thus enabled young males to get a foothold in urban centers and look for nonagricultural opportunities.[11] Adem Aliyu Jaylan, born in 1939, was one such young man who saw a great opportunity in the khat-selling business. He recalls:

> As a young man, I collected *taacharo* (low-grade khat) and sold it in the market in order to buy salt and gasoline. This got me in touch with the business of khat. I earned money selling to khat traders fresh banana leaves used as khat wraps to protect it from heat and kept it fresh. Later, I began to work for khat traders, tending the khat they had bought until

[8] Marc Weill, 'Marketing of Agricultural Products in Harar Highlands, Hararghe Administrative Region, Ethiopia, with Special Reference to Kombolcha *Woreda,*' Department of Agricultural Economics, Alemaya Collage of Agriculture (Alemaya, 1984), pp. 25–6.

[9] Lakew Birke, *A Survey of the Marketing Methods and Facilities in Alemaya District,* College of Agriculture, Haile Sellassie I University (Dire Dawa, 1974), pp. 15–16.

[10] Interview: Hajji Hassan Mohammed, Haramaya, May 5, 1994.

[11] Interviews: Mumed Ahmed Umer, Haramaya, May 17, 1994; Luye Said, Haramaya, May 26, 1994.

Photo 7.1 *The khat market center in AwWoday, April* 1994

enough had been collected. Then, I moved on to processing, i.e. trimming and packaging khat into *aqaaras* weighing approximately 1 kg. For this work, my employer paid me 1 birr/day. My pay increased gradually, from 1 birr to 2 and then, in 1956 [Ethiopian Calendar], to 3 birr.[12]

Alemayehu Woyessa's life history parallels Adem's. While Adem entered the khat business at the production site, Alemayehu joined it on the market side, working for a khat merchant in Dire Dawa. Recalling his arrival in Dire Dawa in 1949, Alemayehu says:

> As soon as I mastered reading and writing, my parents decided I had had enough schooling and that it was time for me to turn to farming. My schoolmate in Deder told me that I could find employment in Addis Ababa and I came to Dire Dawa to proceed to Addis. I stayed in a house owned by an Arab guy. The landlord offered me a job and I stayed on in Dire Dawa. I worked five days a week recording the amount of khat bought by agents of my Arab employer and brought to the processing center. I also transported the processed khat to the airport for export to Aden and collected payments for the khat that was exported the previous day. When a khat exporting company was established [in 1949], those of us who had acquired experience in the business became employees of the Arab members of the new company.

[12] Interview: Adem Aliyu Jaylan, AwWoday, January 19, 2000.

> One Arab who had a receiving agent in Aden employed me as his purchasing agent and sent me to AwWoday. I was paid 5 percent commission based on the amount of export-grade khat I purchased.[13]

By providing nonfarm employment, the rising khat market attracted young males of rural Harerge to flock to the large market centers. The migratory process created other opportunities in agriculture. Farmers were able to increase their plot size by incorporating the share of their siblings who had migrated to urban centers. In the early 1970s the urbanized farmers began taking up other businesses apart from retailing khat, and invested their profits mostly in small shops that carried commodities in demand in the rural areas such as textiles, batteries, salt, cooking oil, sugar, blankets and so forth. A few were able to set up gas stations while others drove taxis between Harer and Dire Dawa and other towns in the region.[14] Alemayehu Woyessa describes AwWoday's rapid transformation, saying:

> When I came to AwWoday in 1942 E.C. (1949), the place was called Didimtu, owing to the color of the soil here. There was only one house, owned by a Somali man called Sherifo. The khat transactions took place under a tree next to Sharifo's house. I bought khat there and turned it over to the company on behalf of my employer. AwWoday grew from such a humble beginning to the most important khat purchasing center in my own life time. Everything in this town was built by the farmers who left their farms and moved here. They built houses, stores, restaurants, gas stations and so forth. AwWoday is a town of farmers who are still moving in and building their residences and businesses.[15]

The farmer-traders retained the right to return to their land, although there is no evidence that anyone reverted to farming. Venturing into nonfarm businesses, most of them severed ties with their rural lands and became traders. Young farmers who moved to small towns on the highlands continued to be active in the khat business but those who moved to Dire Dawa went into new businesses. In the 1960s the growing network of government-built roads facilitated the increased use of motor vehicles for freight and passenger services. Over time, the migrants were able to accumulate capital and branch off to businesses that have little connection with khat, mainly ownership of commercial transport vehicles, gas stations and urban real estate. For such businessmen, returning to the farm was not an option; in fact, they became consumers and purchased khat for their own consumption.[16]

[13] Interview: Alemayehu Woyessa, AwWoday, January19, 2000.

[14] Interview: Mumed Ahmed Umer, Haramaya, May 17, 1994.

[15] Interview: Alemayehu Woyessa, AwWoday, January19, 2000.

[16] Interview: Yusuf Hammo, Haramaya, May 18, 1994. For similar processes of migration, accumulation and mobility, see Sara Berry, *Fathers Work for their Sons: Accumulation, Mobility and Class Formation in an Extended Yoruba Community* (Berkeley, 1985); idem, *No Condition is Permanent: The Social Dynamics of Agrarian Change in Sub-Saharan Africa* (Madison, 1993).

The evolution of the internal marketing structure

The marketing of khat was gradually made efficient by a complex structure of marketing that emerged after the dissolution of CHATHA, the company which, before the slump of the mid-1960s, dominated the export to Aden. Individual exporters came to dominate the business and the new structure involved three types of intermediary buyers: purchasing agents, brokers and independents. The agents worked for the exporters and operated mainly with funds supplied by the latter. Most of the exporters were expatriates, Arabs and Djiboutians, but there were also some Ethiopian Somalis. The agents came to such marketplaces as Haramaya and later AwWoday, but did not actually buy khat directly from producers. They hired brokers to whom they advanced cash for the purchase of khat. The purchased khat was transferred to the agent, who would pay the broker and give him a further advance, a cycle repeated day after day.

By contrast, independent buyers often operated independently of the exporters in the sense that they were not tied down to one firm and received no cash advances. They bought khat with their own capital and sold it to the exporters. Some moneylenders also acted as independent buyers, extending credit to farmers before harvest seasons on the understanding that they would purchase crops at an agreed-upon price.[17]

This marketing system favored traders because it gave them the power to determine prices. Besides the exporters, there were no alternative buyers who could have fostered competition and raised prices. With no serious competition, the exporters were able to dictate prices and take much of the profit from the khat business. Next to the exporters, the purchasing agents and brokers profited most from the business. Studies dealing with agricultural marketing almost invariably inveigh against what was considered to be an unduly large number of middlemen, who prospered by defrauding producers. Both officials and farmers invariably disparaged the middlemen, but they were not able to do business without them. The 'khat brokers,' or middlemen, were an indispensable part of the business; merchants relied on their services in collecting produce from the thousands of small growers, while farmers enjoyed the marketing services and loans (in the form of crop advances) they provided.[18]

Although exporters and traders benefitted the most, the character of the market, specifically the perishability of the commodity and its consequent price volatility, did not allow for the emergence of large-scale enterprises dominated by big merchants. It is probably no exaggeration to state that the lack of technology of preservation and the inadequacy of transport infrastructure kept the khat business fairly 'democratized.' Price speculation or market manipulation was not possible with a commodity whose 'shelf life' was no more than two days. Nor was the

[17] Interviews: Ahmed Bakar, Haramaya, May 20 and 27, 1994; Hajji Mohammed Abdurahman, Dire Dawa, March 26, 1994; Hajji Hassan Abdo, Dire Dawa, March 26, 1994.

[18] Lakew, *A Survey of Marketing Methods*, p. 6.

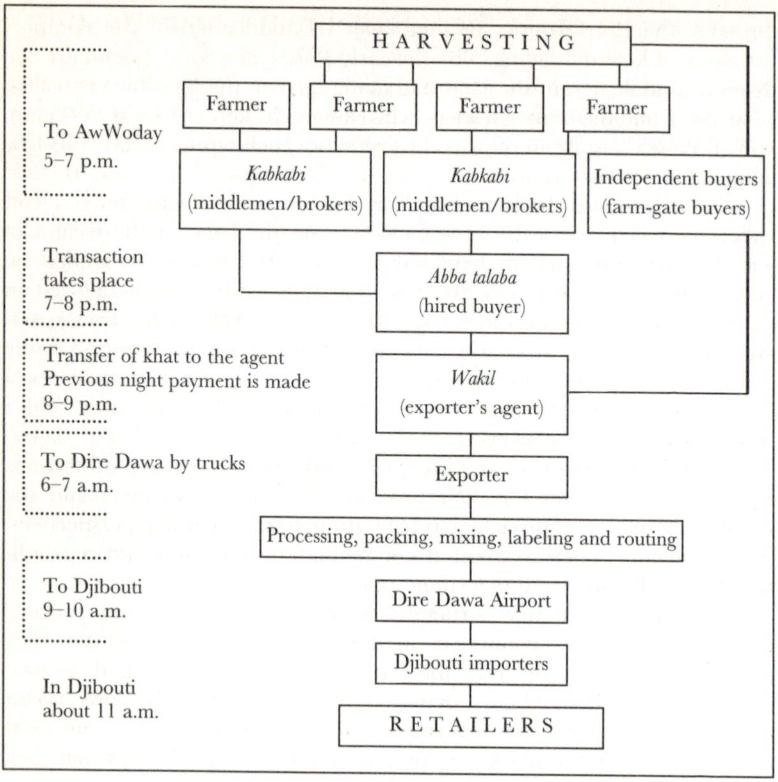

Figure 7.1 *Flow chart of khat marketing*

producer able to predict in which direction prices would turn in the absence of information regarding prevailing market conditions. The only workable marketing strategy was to sell the product as quickly as possible at prices that seemed to yield reasonable profit. A trader could not expect consistency at the marketplace for there was no guarantee that the price offered for a certain kind of khat one day would hold for the next day. Prices were set by the quantity of khat that had reached the marketplace at a particular time. If a producer arrived late, he would have to sell his khat at a lower price, thus forcing prices down, or face a danger of losing it all when his khat wilted.[19]

The marketing system which benefitted traders began to change as farmers who had moved to urban centers increased their involvement in the marketing process. The farmer-traders initially sold khat produced on their family farms, but subsequently became general agents. Since the farmer-traders were more trusted by khat farmers, they successfully inserted themselves between the producers and the agents of big exporters and the

[19] Interviews: Hajji Hassan Ahmed, Haramaya, May 18, 1994; Hajji Hassan Abdo, Haramaya, May 18, 1994; Abdo Shanqo, Haramaya, May 19, 1994.

brokers, thereby forming another layer of middlemen in the complex structure of khat marketing.[20] By the early 1970s there were essentially two types of middlemen in the khat marketing system: the khat buyers, called *abba talaba*, and the farmer-traders, who came to be known as *kabkabi* (Figure 7.1). Both had a stake in the khat business, but each represented conflicting interests in relation to prices.

The *abba talaba* was paid by his employer, the exporter's agent, otherwise called the *wakil*, at a fixed rate on the basis of the weight of the khat he purchased or negotiated to buy. His income depended on their ability to drive prices down so that they had more money left in their hands to purchase more khat. On the other side of the transaction worked the *kabkabi*, who were the principal agents for farmers at the marketplace. They were paid by farmers who sought their services a commission of 5–10 percent of the proceeds of the khat they sold.[21] Since their income depended on the price they negotiated, the *kabkabi* had a vested interest in driving prices up and ensuring that farmers received better prices for their produce. Both the *abba talaba* and the *kabkabi*, therefore, approached price-setting from opposing perspectives. The conflicting interests created competition, negotiation and a condition for market forces to determine price.

Even though the new structure dominated the marketing process, independent buyers continued in the khat business. Because they purchased khat directly from the farmers and operated with their own money, independent buyers bypassed the *kabkabi* and the *abba talaba*, who were responsible for determining daily khat prices. The independents' leverage in the marketing process stemmed from the fact that they purchased futures and collected the produce themselves, sparing the farmer the cost of farm-to-market transportation and the uncertainties of finding a buyer at the marketplace. Independent buyers became most active during dry seasons when khat was scarce, and beat other traders owing to the futures they had purchased.[22]

Buyers for domestic markets often used a different and less complicated mechanism of purchasing khat. The *kabkabis* were the main intermediaries in the domestic system as well, serving farmers as agents. The transaction of khat intended for domestic consumption usually took place in open market centers, where the *kabkabi* transferred the produce to wholesale buyers. There were a number of khat markets in various towns, but the trade centered on Dire Dawa, Haramaya and AwWoday. The wholesale buyers distributed to retailers in the major towns, the majority of whom were women. Farmers' wives sold at a low price poor-quality khat, known as '*faaqa*' or '*taacharo*,' directly to consumers.[23]

[20] Interview: Mumed Ahmed Umer, Haramaya, May 17, 1994.
[21] Interview: Hajji Mume Kiko, Haramaya, May 20, 1994.
[22] Interviews: Yusuf Hammo, Haramaya, May 18, 1994; Mume Ahmed, Kombolcha, June 6, 1994.
[23] Interviews: Ibrahim Dawe, AwWoday, May 30, 1994; Mahdi Abdule, AwWoday, May 30, 1994.

Photo 7.2 *Picking khat. Men harvest the best-quality leaves*

Photo 7.3 *Picking khat. Women get the low-grade* faaqaa *or* taacharo

Photo 7.4 *Women selling khat, AwWoday, June 1994*

Photo 7.5 *Loading khat onto trucks for market or processing, AwWoday, June 1994*

Coffee had been an important cash crop for farming families in this area before the Italian invasion. By the early 1970s it had fallen behind khat. There are no records to show precisely how much khat was sold at the marketplace daily. The fragmentary figures that exist nevertheless underscore the financial significance of khat to farming families. In 1972–3, for instance, 135 farming households sold 40,000 birr worth of farm crops, 59 percent of which was comprised of stimulants, khat and coffee. Khat constituted the greater part of the value of stimulants, amounting to 22,000 birr, or approximately 55 percent of the total value of the crops sold. The figures also show that the entire coffee crop was marketed while only 52 percent of the khat was sold. Despite the differential, coffee accounted for a mere 7 percent of the value fetched by the stimulants. The fact that only 52 percent of the total khat production accounted for 93 percent of the value of the marketed stimulant crops denotes a significant modification in the farming strategy pursued by most farming families. These adjustments included changes in the logic of production that informed the farmers' decision in regard to land use, labor allocation and cropping patterns.[24]

Modifying production strategies

Some studies carried out in the late 1960s suggest that declining markets and unfavorable prices had caused many farmers to abandon their khat fields[25] or halt further planting of new trees.[26] Yet some statistical data do not support the assertion that khat was being abandoned. The evidence indicates that the uncertainties associated with khat marketing affected producers' attitudes and called into question the viability of the farming strategy they had developed to cope with arable land scarcity. Oral information suggests that the most immediate and visible change in the farming system was the conversion of khat fields from the *deema* to the *daalota* variety. Before 1964 the former was preponderant because it was popular in Aden and northern Somalia. Following the collapse of the foreign markets, demand for the *deema* variety declined drastically, and farmers in Haramaya, Kombolcha and Hundene *woredas* uprooted their *deema* khat and replaced it with the *daalota* variety more popular in Djibouti. The Gara Mula'ata–Wobberra region, which joined the khat market in a significant way quite belatedly, continued to produce the *deema* variety for local markets and northern Somalia.[27]

The process of switching from *deema* to *daalota* represented an informed,

[24] Lakew, *A Survey of Marketing Methods*, pp. 13–14.

[25] Telahun Makonnen, *Economic and Sociological Characteristics of Peasant Families at Alemaya, The Chercher Highland*, Department of Agricultural Economics and Business, College of Agriculture, Haile Sellassie I University (Dire Dawa, 1974), p. 3.

[26] Miller and Telahun, *Organization and Operation*, p. 19.

[27] Interviews: Mume Ahmed, Kombolcha, June 8, 1994; Abdulahi Ahmed, Haramaya, May 17, 1994; Amin Usman Mohammed, Haramaya, May 9, 1994.

Photo 7.6 *Khat interplanted with sorghum, sweet potatoes, spice grasses and some maize*

conscious and forward-looking response to changing market situations, though, to the unknowing observer, it might have appeared as an effort to eradicate khat. In fact, the events of the mid-1960s accentuated the vulnerability of the khat market and reinforced farmers' doubts about wholesale adoption of cash crops even when profits are high. Despite the surge in the khat market in the 1950s and early 1960s, food crop production remained the dominant farming activity, and, in the early 1970s, farm land grew extremely scarce. The average family held between 0.5 and 1.5 hectares. The Oromo farmers of the Harerge highlands believed that they could use their land to the best advantage and provide some crop insurance by growing several food crops together.[28] One study, based on research conducted in 1968–9, summarizes Harerge's cropping pattern:

> Mixed cultivation is the general practice employed in the area. On the average about 70 percent of the fields were interplanted with two or three kinds of crops. Single cropping is an exception rather than the rule. It is not uncommon to see as many as five kinds of crops interplanted in a field. The most usual crops that are found interplanted are sorghum, corn, chat, beans and sweet potatoes.[29]

[28] Miller and Telahun, *Organization and Operation,* pp. 18–19.
[29] Telahun, *Economic and Sociological Characteristics,* p. 8.

The planting of diverse crops indicates the premium placed on food self-sufficiency. Sorghum occupied 33 to 53 percent of most families' farm land, followed by corn and khat. A sizeable area of the crop land was also devoted to growing sweet potatoes, a few Irish potatoes, onions and vegetables. Most surveys of the consumption patterns of Harerge's farming households show that almost all their sorghum and corn and two-thirds of the khat were consumed on the farm.[30]

The finding of a 1965 study of two farming communities reveals the direction of change in the agrarian system of the Harerge highlands in the early 1960s. Most of the farming households maintained khat orchards in their cropping system, usually intercropped with other food crops, with the orchards occupying the largest percentage of land on smaller holdings, those with land areas of less than 0.50 hectares. Larger holdings of approximately 4.50–4.99 hectares maintained exclusive khat orchards but devoted more land to the production of cereals. The common theme in both large and small holdings' cropping systems is that most farming families subscribed to the notion of maintaining food self-sufficiency and avoiding monoculture.[31] Diversification was adopted as an insurance against land scarcity, disease and pest infestations, but it was also evident that, the smaller one's holding, the greater the tendency to increase the production of khat. Cash was needed to cover the household's expenditure on capital goods (tools, seeds, livestock), consumer products (fuel, clothing, household utensils) and taxes.[32] Khat provided sufficient funds to cover these expenses and for some savings. As part of a diversified cropping system, khat not only fitted well into the environment and the agricultural cycle of small-farm households, but also provided a regular source of income to meet household expenditure with a comparatively low labor investment.[33] Khat also consistently maintained a price advantage over other crops. For example, it yielded ten times the return of coffee.[34] Khat's gross return per hectare ranged from $1,163 to $3,900[35] as compared to sorghum's $264 on average.[36]

Despite the impressive return, khat prices were subject to severe seasonal fluctuation, the result not only of the logic of demand and supply but also of the nature of the commodity, the availability of rains and occasional government intervention in the market. In 1964, for example,

[30] K.C. Davis, Ahmed Mohammed and W.A. Wayt, *Farm Organization of Terre and Galmo Villages: Harer Province, Ethiopia*, Imperial Ethiopian College of Agriculture and Mechanical Arts, Experiment Station Bulletin 42 (Addis Ababa, 1965), p. 23.

[31] Ibid., pp. 12–16.

[32] Ibid., pp. 25–6.

[33] Interviews: Mumed Ahmed Umer, Haramaya, May 17, 1994; Yusuf Hammo, Haramaya, May 18, 1994; Mume Ahmed, Kombolcha, June 8, 1994. Leonard F. Miller, *Input–Output Data for Chat and Sorghum Production, Harer Highlands, Ethiopia*, Imperial Ethiopian College of Agriculture and Mechanical Arts (Alemaya, 1965). Quoted in Quenemoen, Watts and Mosley, *Economics of Irrigation*, pp. 42–3.

[34] Amare and Krikorian, 'Coffee's Rival,' p. 370.

[35] Quenemoen, Watts and Mosley, *Economics of Irrigation*, p. 25.

[36] Demissie, *The Kottu of Harerge*, p. 89; Quenemoen, Watts and Mosley, *Economics of Irrigation*, p. 25; Lakew, *A Survey of Marketing Methods*, p. 14.

when the government restricted exports to northern Somalia, prices plummeted, demonstrating that, while potential incomes were high, there were significant elements of risk in expanding the cultivation of khat. Not only were prices volatile but the market itself was unstable and unreliable. Market unpredictability was one reason for the primacy of food self-sufficiency in farming strategies.

Khat's relative profitability should have made other cash crops unattractive and unsuccessful. Nevertheless, in an effort to even out price fluctuation and retain more value from their own labor, producers diversified the cash crops they cultivated on their farms. From the mid-1960s on, there was increasing cultivation of vegetables, particularly of potato, especially in areas where irrigation was possible. While the use of electric water pumps facilitated irrigation, the opening of the Djibouti market for Harerge's vegetables created demand. In the late 1960s most farmers in Haramaya's Yatu valley and in Kombolcha's Faldhaana region had increased potato cultivation on their farms through irrigation from such streams as Lega Hida and Lega Jeneta.[37] Although Harerge was the nearest source of supply, Djibouti imported vegetables from Eritrea and Europe because Harerge's vegetables did not meet its quality standards. In 1963 Haramaya's farmers established the Farmers' Marketing Cooperative to ensure quality and began exporting vegetables to Djibouti.[38]

By the late 1960s vegetable production was the second highest cash earner for farming households, a logical response to the uncertainties of the khat market. Vegetables, particularly potatoes, enjoyed a much longer 'shelf life' than khat. If market prices happened to be low, the tubers could remain in the ground for a few weeks. In this way, potato provided income the whole year around because it could be harvested at any time of the year if irrigation was available. In providing a continuous cash flow, potato possessed the characteristic farmers liked about khat.[39] Potatoes also were important as an alternative food crop during times of acute food shortages. Planted in April, short-cycle varieties matured in July, before the first green maize crops were ready to be harvested. The crop remained in the soil from August to May, providing badly needed food during a period when the grain stocks are low. Put another way, both sweet and Irish potatoes provided a food bridge between the two annual cereal harvests.[40]

The increasing importance of potato as a cash crop did not presage the end of khat expansion. Land-use figures at least suggest that the expansion of khat continued, despite all the difficulties the export market encountered throughout the 1960s. In the early 1970s, 26,800 hectares

[37] Lemma, 'A Regional Study of Hara Maya,' p. 56.

[38] V.F. Aman, *Organization and Operation of a Marketing Cooperative in Ethiopia*, Imperial Ethiopian College of Agriculture and Mechanical Arts, Experiment Station Bulletin 43 (Addis Ababa, 1965), pp. 3–5.

[39] Lemma, 'A Regional Study of Hara Maya,' p. 56; Wibeaux, *Agriculture in Hararghe*, p. 50.

[40] Wibeaux, *Agriculture in Hararghe*, pp. 51–5.

Photo 7.7 *Replacing food crops with a cash crop: planting new khat trees into a maize-cropped field, Haramaya, March* 1994

were devoted to khat in Harerge, a phenomenal growth compared to the 7,009 hectares given for 1961. The land-use pattern in the various *awrajas* reveals the relationship between khat cultivation and land shortages as well as the proximity of farm areas to transport. In Chercher, where population pressure was not so much of a problem and transportation had not been available until 1963, coffee remained the predominant perennial crop, accounting for 11 percent of the total cultivated area, while khat represented only 3 percent. Sorghum and maize accounted for 65 percent of the total area under cultivation.[41] In Harer Zuria, the sub-province with the highest population density and better transportation facilities, khat occupied 13 percent of the total area under cultivation compared to coffee's 3 percent and 81 percent of the area under perennial crops compared to coffee's 19 percent. Sorghum and maize constituted 79 percent of the total area under cultivation, suggesting that food crop production remained important.[42]

[41] The figure for khat was given by Amare and Krikorian. The authors provided this figure for Ethiopia as a whole, not specifically in reference to Harerge. Since they conducted their research in Harerge, I presume that the hectarage referred to the areas of their research. What casts more doubt on the numbers is not only their extremely low estimate but also that Amare and Krikorian relied on T. Hagos, *Chat in Ethiopian Economy,* a document I was not able to locate in the major libraries and research centers in Ethiopia to verify the estimate.

[42] Central Statistical Office, *Results of the National Sample Survey, Second Round: Land Area and Utilization,* 5 (Addis Ababa, 1975), p. 77.

The land-use statistics indicate that the disappearance of an export market did not affect the overall trend in the direction of expanding khat cultivation. The growth indicates that demand did not slacken enough to warrant suspending or abandoning the production of khat, despite the loss of foreign markets. Production continued to expand in response to the increasing demand generated by the contraband market and a substantial rise in domestic consumption.

Conclusion

Farmers do not necessarily watch market conditions in determining farming strategies. The security – or the subsistence – imperative of the family comes first when farmers make decisions about selecting crops, allocating land and deciding how much of the produce to sell. In Harerge food crops were generally not marketed. More than 95 percent of the grains, leguminous and root crops were produced for household consumption. Harerge's farmers increased the portion of their plots devoted to khat at the expense of food crops to meet their cash requirements rather than quickly reaping the financial rewards of an ever-increasing demand. The fact that food crop production remained dominant, despite the increasing demand for khat, points to the limited power of the market in dictating farming strategies and production levels.

Until the mid-1970s, the expansion of khat cultivation in Harerge was not quite at the expense solely of food crop production. The only crop that khat apparently displaced was coffee. Informants relate how they and their parents cut down coffee trees and replaced them with khat.[43] When the market for the commodity was expanding in the 1950s and early 1960s, khat cultivation was extended to marginal lands. From the mid-1960s on, khat cultivation proceeded with emphasis on interplanting the shrub with other crops. The logic of production in the 1960s and early 1970s was based on how best to use the resources available in the face of arable land shortages, uncertain markets for cash crops and increased government intervention in order to meet the subsistence needs of the family and obligations to government.

[43] Interviews: Nuriya Sheik Ibrahim, Haramaya, May 6, 1994; Ismael Mohammed, Kombolcha, June 7, 1994.

Part Three

Khat & Agrarian Transformations

Eight

The Khat Boom

Production & Marketing in a Liberalized economy

1974–91

[illegible]

[illegible]

[illegible]

Eight

The Khat Boom
Production & marketing in a controlled economy 1974–91

> This khat tree is the soul of the entire farming community in Harerge. Khat is for us what the tail is for the bee, a seat of its soul. No one can live without a soul and we cannot survive without khat. If khat is eradicated [say, by someone's order], alas, all the inhabitants of this land will perish at once.
>
> – Hajji Abdi Bakar, 1994

Before the Ethiopian revolution of 1974, khat's commoditization was initiated and sustained by a convergence of events and processes. In response to population pressure farming families introduced new crop varieties, intensive cultivation techniques and diversification. They also took advantage of improved road networks and other infrastructural developments and increased the production of high-value cash crops. The internal (demographic pressure) and external (infrastructural improvements) factors together produced a dynamic force of socioeconomic change, which led to a profound transformation in Harerge's agrarian system. From the mid-1970s on, the process of agricultural transformation was accelerated by the changes in agricultural production and marketing policies that were introduced following the overthrow of Haile Sellassie's government. Under the imperial regime, land and access were concentrated in the hands of landlords, and marketing and pricing of agricultural commodities were generally left to the market. The post-imperial government, however, abolished the privileges of the gentry and increased the level of state intervention in the marketing and pricing of agricultural commodities.

In the decade that followed the revolution, the production and marketing of khat assumed new patterns and significance. In 1975 khat constituted 6.6 percent of the total land under cultivation in Harerge. In less than a decade, this proportion doubled and, by 1983, khat was credited with 13 percent or 96,445 hectares of the cultivated land area. The volume of domestically marketed khat rose from around 1,300 tons in the early

1960s to an estimated 4,800 tons in the late 1980s.[1] Foreign-exchange earnings from khat, which had hovered around 5 million birr to the mid-1970s, reached 32 million birr in 1984 and peaked at 36 million in 1989. In the early 1980s revenues from income taxes averaged 23 million birr per year.[2] Some experts have suggested that, by the mid-1980s, khat was the second largest Ethiopian export, after coffee.[3]

The dramatic change in the status of khat as a commodity may be ascribed to the implementation of the policy of land redistribution, which gave farmers plots of their own. Most farming families subsequently began growing khat, preferring it over other cash crops. The farmers' preference is partially owed to the marketing policies of the Provisional Military Administrative Council (PMAC), specifically the stricter control imposed over coffee marketing and pricing, which in turn encouraged farmers to shift to the then less regulated khat. As soon as the profitability of khat became evident, however, the government extended its involvement in khat marketing after 1979 by creating a government-run khat export enterprise. In spite of active government support, the company eventually failed to compete with private traders, thus leaving much of Ethiopia's khat export in private hands.

Land redistribution, khat production & agricultural productivity

In February 1975 the PMAC proclaimed a land reform law, which eventually abolished tenancy, revoked the titles and rights of landlords and provided for the redistribution of land among farming households. Peasant Associations (PAs) were formed and entrusted with the task of implementing the law, being given considerable latitude in deciding the specific procedures of distributing land. The PAs followed a variety of methods, which resulted in local variations in the size and quality of plots. Overall, however, most farming households were given usufructuary rights over the land they cultivated, although in some cases, the holdings were insufficient to sustain a household. The new system comprised predominantly user-operated small farms and a few cooperatives with a PA administering collectively owned lands.[4]

[1] Income tax revenues were 2,600,000 birr and 9,462,529 birr in 1963 and the 1987/8 fiscal year, respectively. Calculation on the basis of income-tax rate, which was 0.50 birr/kg, will give us the estimated amount of officially transacted khat that is shown in the text.

[2] Eastern Ethiopia Zone Planning Office, *Bulletin of the Socio-Economic Conditions in the Zone* (Harer, 1988), p. 8 and Appendix 9 and 10; Ministry of Agriculture, *General Agricultural Survey Preliminary Report, 1983–84 (1983) Harerge*, vol. 2 (Addis Ababa, 1984); Customs and Excise Authority, Annual External Trade Statistics, 1960–1991.

[3] Christopher Clapham, *Transformation and Continuity in Revolutionary Ethiopia* (Cambridge, 1988), p. 186.

[4] Desalegn Rahmato, *Agrarian Reform in Ethiopia* (Uppsala, 1984), pp. 37–44; Kidane Mengisteab, *Ethiopia: Failure of Land Reform and Agricultural Crisis* (New York, 1990), pp. 97–107. For local distribution analysis, see Mulugeta Gashaw and Minilik Tibebe Sellassie, *Ethiopian Village Studies, Hadele Keke, Kersa Woreda, Hareghe* (Harerghe, 1995), pp. 6–7.

In Harerge province the longstanding problem of arable land scarcity was eased but not eliminated by forest clearing, reduction of fallow periods and the incorporation of marginal lands. Redistribution under the *derg* did not provide a lasting remedy for arable land shortage, but the ending of tenancy removed structural constraints on the region's agriculture in the sense that farmers were freed from sharecropping arrangements that favored the production of grains over tree crops and were empowered to make decisions regarding the cropping patterns on their individual plots. The fact that land reform did not solve the main problem of Harerge's agriculture stimulated farmers to pursue new strategies of raising productivity. One farmer informant from Faldhaana, Kombolcha, summed up the new dynamics of agrarian change in Harerge quite succinctly: 'one who possesses large tracts of land, relies on his land to produce enough for his family's sustenance. One who has less land, innovates to overcome his predicament.'[5]

The redistribution of land increased cultivation areas, eased population pressure and provided capital for improving productivity. Many farming families took up the production of marketable commodities by using irrigation to increase productivity and the levels of dry-season production. However, not every area in Harerge was endowed with water resources, and there was not enough irrigable land to distribute equally among all farmers. Not surprisingly, the distribution of the available irrigable land favored PA officials and wealthy farmers,[6] inevitably producing disparities and social conflicts. PAs sought to reduce major conflicts by scheduling and rationing water and by encouraging intercropping.[7] In the 1980s, after more than a decade of effort, the highest proportion of areas irrigated was just under 25 percent of the arable lands.[8]

Farmers dealt with water scarcity by building new reservoirs, collection basins and canals. Some farmers used whatever profits they made from the irrigated crops to purchase pumps for their own use and to rent out. In that sense, it is possible to say mechanical irrigation revolutionized the use and management of land and water, the scarcest resources of the agrarian system.[9] By the early 1980s most farming families in Harerge pursued irrigation as an important strategy for increasing agricultural productivity.

Prior to the Ethiopian revolution these innovations were not commonplace. Landlords owned much of the irrigable land and sold water to farmers from their storage basins (*kuré*). In Kombolcha, for instance, the cost for a single watering was between 25 and 30 birr, depending on the size of one's plot.[10] After 1975 farmers expanded both their arable and irrigable lands by appropriating lands and *kurés* previously owned by

[5] Interview: Mume Ahmed, Faldhaana, June 8, 1994.

[6] Bernadette Risoud, *Creation and Development of a Producer Cooperative: Innovations and Interactions: The Case-Study of Tula in Harerge Highlands*, Farming Systems Research, Alemaya University of Agriculture (Alemaya, 1988), p. 11.

[7] Wibeaux, *Agriculture in the Highlands of Hararghe*, p. 54.

[8] Harmen Storck *et al.*, *Farming Systems and Farm Management Practices of Smallholders in the Hararghe Highlands* (Kiel, 1991), pp. 16–17, 63–4.

[9] Quenomoen, Watts and Mosley, *The Economics of Irrigation*, p. 25.

[10] Risoud, *Evolution of Peasant Agriculture*, p. 22.

Table 8.1 *Changes in the share of irrigable land & khat cultivated areas in Tula: Sample of six farming households in hectares*

Farm	Arable land 1940s	Arable land 1987	Irrigable land 1940s	Irrigable land 1987	Khat area 1940s	Khat area 1987
A	0.42	0.42	0.25	0.35	0.35	0.45
B	2.16	2.81	0.15	0.75	0.69	0.95
C	2.02	2.25	0.00	0.11	1.12	1.12
D	1.46	2.10	0.32	0.77	0.20	0.84
E	1.62	2.20	0.60	0.60	0.30	0.30
F	0.50	0.50	0.15	0.00	0.35	0.40
Ave.	1.36	1.71	0.25	0.43	0.50	0.68

Source: Risoud, *Evolution of Peasant Agriculture*, p. 23.

Table 8.2 *Changes in the share of irrigable land & khat cultivated areas in Kako:*[11] *Sample of eleven farming households in hectares*

Farm	Arable land 1940s	Arable land 1987	Irrigable land 1940s	Irrigable land 1987	Khat area 1940s	Khat area 1987
A	0.61	0.71	0.05	0.14	0.12	0.29
B	1.17	1.48	0.29	0.33	0.15	0.21
C	0.60	1.10	0.00	0.00	0.00	0.21
D	0.80	1.29	0.00	0.00	0.05	0.25
E	0.91	1.00	0.00	0.00	0.05	0.35
F	1.10	1.10	0.00	0.03	0.21	0.31
G	1.80	1.00	0.00	0.43	0.21	0.21
H	0.86	1.03	0.12	0.17	0.09	0.28
I	0.89	1.30	0.00	0.00	0.23	0.00
J	0.44	0.82	0.00	0.44	0.15	0.25
K	1.34	1.39	0.57	1.07	0.72	0.72
Ave.	0.96	1.20	0.09	0.24	0.18	0.28

Source: Risoud, *Evolution of Peasant Agriculture*, p. 24.

landlords and by employing innovative ways of using the available water resources. According to the findings of one study, farmers residing in Tula and Kako increased their irrigable lands by nearly 119 percent in less than twenty years. The increase was less in Tula, where water resources were poorer, amounting to 72 percent, while, in Kako, it was a remarkable 167

[11] The two tables on the expansion of arable and irrigable lands show change over the last forty to fifty years. The average age of the heads of households that were studied was 65. Assuming each farmer established a family at the age of 20–25, the time of inheritance for all households would be forty to fifty years ago.

percent. By 1986, irrigable areas in these two localities represented an estimated 30 percent of the total cultivated land (Tables 8.1 and 8.2).[12]

For farmers whose lands were located near large lakes such as Haramaya, Addele and Finkile, irrigation had always been an important technique in the production of khat and vegetables.[13] It is important to emphasize that the increase in irrigable lands shown in the tables occurred in districts where dry-season production was insignificant before 1975. Until the mid-1960s cash crops played a supplementary role in the agrarian system, since most farmers were involved primarily in rain-fed grain production for food self-sufficiency. The cropping pattern was to plant up to 90 percent of the land with grains, using intercropping and multicropping methods. Nongrain crops such as sweet potatoes, fenugreek and beans were grown on small sections of individual farms; but these, like grains, were stored for consumption rather than for market. Khat was the most popular of the cash crops, preferred probably because of the existence of more advantageous production techniques and a sophisticated marketing system.[14] Although after the Second World War, the importance of khat increased steadily, the proportion of land devoted to its cultivation surpassed 10 percent of the total cultivated lands only in the early 1980s.[15]

The significance of khat lay not so much in the size of the land it occupied but in its productivity. Farmers optimized their return, especially during dry seasons when khat prices usually rose to ten times the rainy season values. The use of irrigation intensified khat production by extending harvest periods and increasing yield per unit. In areas where irrigation was not possible, farmers increased khat production by putting more land under cultivation, often at the expense of other crops. As Tables 8.1 and 8.2 show, the land area devoted to growing khat increased at a much faster rate than the total land area brought under cultivation. In fact, the growth in the amount of irrigated land allocated to khat production was made possible by increases in the amount of arable lands.[16]

In the early 1970s irrigation became even more important following the introduction of a new potato cultivar by the Ministry of Agriculture. As both food and cash crop, the potato quickly became a strong competitor with khat for irrigable land, labor, water and fertilizer. In the early 1980s some farmers were able to obtain an annual return of as much as 12,000 birr per hectare from potato. This level of productivity was rare and was attained only by a few farmers who had permanent access to irrigation. The average gross return for most farmers was 7,600 birr per hectare. During

[12] Storck *et al.*, *Farming Systems in Hararghe Highlands*, p. 63.

[13] Lakew Birke, *A Credit Study of 100 Members of Alemaya Farmers Cooperative Society, 1965,* Imperial Ethiopian College of Agriculture and Mechanical Arts, Experiment Station Bulletin 46 (Dire Dawa, 1966), p. 7.

[14] E. Westphal, *Agricultural Systems in Ethiopia* (Wageningen, 1975), pp. 108–18.

[15] MOA, *General Agricultural Survey*. Another study puts the proportion of cultivated land under khat at 17%. Desta Degefu and Masresha Yimer, *Conditions of Social Security in Harerge* (Amharic), prepared for Eastern Ethiopia Zone Planning Office (Harer, 1986), p. 30.

[16] Risoud, *Evolution of Peasant Agriculture*, pp. 22, 33.

the same period there were reports of a gross return from khat of 6,500 birr per hectare in Kombolcha *woreda* and a higher 9,600 birr for the entire Harer Zuria *awraja*.[17] Considering the actual sales of 50-60 percent of total production, farmers commonly netted 4,500 birr per hectare from khat. In terms of return on investment per hectare, therefore, potato was more profitable than khat.[18]

Despite potato's apparent profitability, most farmers remained interested in khat. First, the lack of capital for investment in seeds and fertilizer made high-input potato management impossible. Although it is true that Harerge's farmers possessed more cash than farmers in other parts of Ethiopia, the amount of cash they were able to allocate for productivity investments was limited. High-input potato cultivation could take up as much as the entire annual cash income of a farming family. Attractive as the profit from potato might have seemed, farmers were not willing to embark on such an expensive investment strategy. Second, investment in potato was extremely risky, owing to the crop's susceptibility to weather shifts and blight. The chance of crop failure from blight was lower during dry seasons and it could be eliminated through increased use of irrigation. However, irrigation was not as readily available as was needed to maintain dry-season potato production. Third, potato required watering every eight days, and the available resources in Harerge highlands were not sufficient to meet such a high demand. Only a few farmers, PA leaders or those having high social status had consistent access to irrigation water.[19] Aware of the cost of feuding over scarce water supply, farmers took little interest in extensive potato cultivation, especially of the improved variety (large potato), which was extremely sensitive to water scarcity.[20]

In contrast, khat offered two cultivation advantages. First, it was more drought-resistant, requiring water every 30 to 35 days. During dry seasons, khat leaves were pruned and the trees kept dormant so as to restart vegetative growth with the first rains. Ideally, two irrigations were necessary to make the trees sprout, but an adequate level of harvest was gained even with a small amount of water. Sometimes, however, farmers waited to bring their khat into production at the time of high market prices. With khat cultivation, farmers were able to make maximum use of scarce water to minimize the social conflicts that were frequent during dry seasons.[21]

Second, khat needed relatively little fertilizer and labor. Chemical fertilizers were generally avoided because they led to crop etiolation. Cultivators mostly relied on the use of animal manure and household waste. In terms of labor input, since more than 70 percent of the khat in the

[17] D. Aehling-Kriesel and Kifle Lemma, *Farming Systems Development Survey Report*, Ministry of Agriculture, People's Democratic Republic of Ethiopia, Field Document TCP/ETH/6659 (1989).

[18] Wibeaux, *Agriculture in the Highlands of Hararghe*, p. 64.

[19] Risoud, *Creation and Development*, pp. 10–11. The Producers' Cooperatives (PC) were empowered to evict nonmember farmers if the PC needed the individual farm. In fact, the PC leaders appropriated fertile and irrigable lands from nonmembers for themselves.

[20] Wibeaux, *Agriculture in the Highlands of Hararghe*, p. 50–4.

[21] Interview: Mume Ahmed, Faldhaana, June 8, 1994.

Photo 8.1 *Khat intercropped with sweet potatoes*

Harerge highlands was grown in association with food crops, the shrub benefitted from the labor and technical investments made on the other crops. During rainy seasons when heavy farm work occurred and khat prices were low, farmers consumed the bulk of khat and sold relatively small amounts. During dry seasons, they raised the proportion of the marketed khat as high as 60 percent of the total production because of favorable prices. Moreover, since, in dry seasons, the intensity of agricultural work decreases, farmers chewed less and sold more khat.[22]

Potato cultivation in association with khat continued on a significant scale for two reasons. First, under the PAs' irrigation water management program, sole-cropped khat fields were ineligible for irrigation water. In other words, irrigating khat was contingent upon the production of food crops. In order to maintain their eligibility for irrigation water, farmers had to cultivate khat in association with food crops, and potato was the preferred food crop to be interplanted with khat. Second, potato was used as both a food and a cash crop. At low-input level, the benefit of cultivating potato lay in its short maturation cycle. It ripens more quickly than grain crops – three months rather than seven to ten in the latter case. When investments were made in seeds and fertilizer, potato gave the highest return in the shortest period of time, particularly on irrigated farms where the potential return was higher and the risk of failure lower.[23]

Indeed, the PAs required khat to be intercropped in order to increase

[22] Wibeaux, *Agriculture in the Highlands of Hararghe*, pp. 42–9.
[23] Ibid., pp. 50–5.

Table 8.3 *Crop production per hectare*

Crop		Yields (quintal/ha)	Prices (birr/kg)[1]	Gross return (birr/ha)
Cereal	low :	9	0.6	550
	high :	25	1.5	3,750
	average :	19	1.0	1,900
Low input potato	low :	—	0.5	2,250
	high :	—	1.0	4,500
	average :	45	0.75	3,350
High input potato	low:	—	0.5	7,600
	high:	—	1.0	17,600
	average :	200	0.75	12,000
Khat[2]	low :	3.5	0.5	200
	high :	7.5	30.0	22,500
	average :	5.0	8.6	4,300

Source: Wibeaux, *Agriculture in the Highlands of Hareghe*, p. 62. (1) Prices are the purchasing prices of food crops in 1986. The extreme volatility reflects seasonal variation. (2) Yields are based on measures taken from intercropped farms. The value of the intercropped crop is therefore added. Since khat could be sold throughout the year, the minimum/maximum gross returns do not reflect actual estimates, but assumptions based on seasonal fluctuations.

food-crop production, but farmers complied for their own reasons. Intercropping with khat allowed for the complementary use of both water and nutrients in space usually characterized by differential root depth and spread. It also reduced the impact of adverse conditions on the ecosystem by providing shelter, conserving soil moisture and decreasing soil erosion.[24] More importantly, intercropping was an insurance technique. Planted together with preferred food crops, khat ensured that some profit was returned from the effort put into the preparation of the land. In 1986 a farmer who interplanted khat with cereals was able to produce on average 4,300 birr from khat sales and some 12 quintals of maize or sorghum, worth 1,200 birr per hectare (Table 8.3). After 10 percent was paid to the *kabkabi* for his brokerage, the farmer took home 3,900 birr in cash.[25] Based on the average grain price of 1.00 birr per kilogram, if a producer purchased grain with the proceeds from khat harvested from one hectare, he or she could buy 51 quintals or grain worth 5,100 birr. If the same land were to be planted entirely with sorghum or maize, the farmer could expect to get a maximum yield of 19 quintals or 1,900 birr.

[24] Addis Tiruneh, 'Gender Issues in Agroforestry,' Paper Prepared for the Workshop 'Land Tenure and Land Policy in Ethiopia,' Institute of Development Studies (Addis Ababa, 1994), p. 6.

[25] In recent years, the price differential with potato from high-input management farms has widened in favor of khat, with per hectare profit for 1994, the year I conducted research in the area, exceeding potato value by over twenty times. The collapse of the post-imperial regime in 1991 created an opportunity for the expansion of the unofficial export market.

The increasing importance of khat is an example of the integration of the region's agriculture into the market economy. In the late 1980s khat prices remained high, most probably owing to foreign demand, but also to the fact that farmers obtained higher land productivity by growing the shrub than by cultivating food crops. Describing why farmers preferred growing khat, rather than grains, one informant explained:

> AB: We are alive today, thanks to khat. If khat wasn't here, we would have all died of hunger, deprivation and lice [disease].
>
> EG: Do you mean that growing khat is better than cultivating grains, vegetables and beans?
>
> AB: Yes! We survived the affliction brought on us by drought, disease and land scarcity because we cultivated khat. Khat is like the tail of a honeybee for us.
>
> EG: What do you mean?
>
> AB: Let me ask you this: what part of a honeybee's body is the seat of its soul?
>
> EG: I don't know, perhaps its wings.
>
> AB: Ohhh, no. Don't you know where the seat of a honeybee's soul is? It is its tail. If that part of its body got damaged, the honeybee is dead.
>
> EG: Alright, explain your point to me.
>
> AB: This khat tree is the soul of the entire farming community in Harerge. Khat is for us what the tail is for the bee. No one can live without a soul and we cannot survive without khat. If khat is eradicated [say, by someone's order], alas, all the inhabitants of this land will perish at once. No one really cares about sorghum or even vegetables anymore. These days people eat rice, pasta and chicken.
>
> EG: Where do you get rice and pasta? Do you buy them in the town markets?
>
> AB: Yes, but, you know, the town is here, the countryside; the countryside has become the modern town. The old ones are ancient relics. We have stores here in the countryside. There are farmers who have trucks to bring stuff here. Why do we need to go to the old towns?
>
> EG: Weren't you, as farmers, supposed to get your food from your grain silos?
>
> AB: You are right. And that used to be the case. But things have changed. Khat has changed everything.[26]

The high returns on investment and the perception of farmers described above indicate a marked change in the logic of production that informed farmers' decisions in regard to cropping patterns and the

[26] Interview: Hajji Abdi Bakar, April 6, 1994, Ganda Lucha, Haramaya. Many of my informants have similar stories, but the informant quoted here has an original way of explaining to me the agricultural situation and farmers' strategies.

use of the available resources for production purposes. The logic of food self-sufficiency, which underpinned farmers' production strategies in the 1960s and 1970s, lost its significance after the 1984 drought. Foreign food assistance was intended for drought victims, but items such as rice, pasta and cooking oil were put up for sale in local markets.[27] When assistance ended, contraband traders took up the task of supplying the food items that were in demand.[28] In addition, the *derg*'s grain marketing policies, which were designed to make grains available for city-dwellers at lower prices, had no mechanism for preventing farmers from purchasing them. The Agricultural Marketing Corporation's (AMC) prices were, of course, higher than the farmers' production costs but far lower than the return from khat sales. As long as food was available in the market, farmers were willing to commercialize their smallholder agriculture. Before the Ethiopian revolution the logic of production was to maintain self-sufficiency while progressively increasing cash crop production without jeopardizing food crop production. The new logic of production of the 1980s was to increase cash crop production as long as there was sufficient food in the market and the crops yielded sufficient cash.[29] In both cases, the food security of the farming household was paramount. Even the commitment to cash crop production was premised on the basic assumption of maintaining a secure source of food.

There are disadvantages to greater dependence on cash crops. Khat prices fluctuate widely and quickly owing to the law of supply and demand. Rain-fed khat mostly comes into production during rainy seasons in the whole region and must be sold at a low price. The level of income from khat, therefore, depends on the possibility of irrigation in order to obtain production at periods of low supply. The scarcity of water for irrigation makes it impossible for all farmers to succeed in dry-season production. Besides, it is impossible to predict what the future of khat will be. The government's attitude has been ambivalent; it condemns khat as a social scourge that must be destroyed,[30] while it continues to collect export duties and taxes on khat.[31] Moreover, Djibouti and Somalia alone may not present a sustainable foreign demand for khat. At this juncture, however, as my informant Hajji Abdi Bakar remarked, the khat

[27] Interviews: Abdulahi Shebo, Ganda Lucha, Haramaya, May 25, 1994; Abdalla Kassim, Ganda Kollobo, Kombolcha, June 8, 1994; Ahmed Abdullahi, Ganda Saqa Sharifa, Kombolcha, June 7, 1994.

[28] During my research in 1994, Harer was flooded with packed foods that were dispatched to the Middle East during the Gulf War for the United States military. The military supplies were reportedly brought to Harerge by contraband traders whose operations extended to Saudi Arabia. Expatriates who lived in Harerge were able to buy cheese, crackers and biscuits that were otherwise unavailable there.

[29] Interviews: Mumed Ahmed Umer, Ganda Issa, Haramaya, May 17, 1994; Abdulahi Shebo, Ganda Lucha, Haramaya May 25, 1994; Beker Tahir, Ganda Sido, Kombolcha, June 6, 1994; Ismael Mohammed, Ganda Kollobo, Kombolcha, June 7, 1994; Sufian Abdulahi, Ganda Lucha, Haramaya, May 8, 1994.

[30] Desta and Masresha, *Conditions of Social Security in Harerge*, pp. 33–6.

[31] Weill, 'Marketing Agricultural Products,' p. 30; Storck *et al.*, *Farming Systems in Hararghe*, pp. 101–2.

phenomenon explains how the commercialization of agriculture sustained the region's demographic weight of more than 400 inhabitants per square kilometer.[32]

Marketing agricultural commodities in a controlled economy

After seizing power in 1974 the military government charted a 'socialist path' for Ethiopia and implemented policies that placed the marketing and pricing of agricultural commodities under tightly controlled government agencies. The *derg*'s policies contrasted with those of the imperial regime, whose involvement in the economy was limited to establishing marketing standards and procedures, quality control and, at times, protecting producer prices.[33] Perhaps the most important policy initiative to affect the khat industry was the government's endeavor to assume greater control over the production, marketing and pricing of coffee, khat's main competitor for land and labor.

The *derg*'s ideological commitment to socialist marketing principles was an important factor, but events also played a significant role in the evolution of the post-1974 coffee policy. In 1975 a severe frost destroyed 50 percent of Brazil's coffee, thereby providing windfall gains for sellers of Ethiopia's beans, who had a tradition of doing well whenever Brazilian coffee fails. In the following years coffee prices remained favorably high owing to the continued decline in Brazil's market share.[34] The Ethiopian government sought to obtain greater foreign-currency earnings by augmenting coffee production. The intention of the government was aptly captured in a letter written in 1977 by a high-ranking official:

> … Ethiopia produces about 200,000 tons of coffee annually. However, only a fraction of that amount is exported. This is primarily due to the lack of coordination and control in the process of transferring coffee from the areas of production to world markets. These inefficiencies of the imperial regime should never be carried over to the new socialist Ethiopia. The Council of Ministers has decided to export at least half of the produced amount, that is, 100,000 tons a year.[35]

Owing to a combination of political and environmental factors, however, coffee production did not keep abreast of international demand.

[32] Milkessa, 'An Economic Analysis', p. 39; Addis, 'Gender Issues in Agroforestry,' p. 4.

[33] Befekadu Degefe and Tesfaye Tafesse, 'The Marketing and Pricing of Agricultural Products in Ethiopia,' in *Ethiopia: Rural Development Options,* edited by Siegfried Pausewang, Fantu Cheru, Stefan Brune and Eshetu Chole (London, 1990), pp. 112–20.

[34] Keith Griffin (ed.), *The Economy of Ethiopia* (London, 1992), pp. 78–80.

[35] MOI-RHP.1, Minister of Trade Ashagre Yigletu to the Ministry of Interior, January 5, 1977.

The government decided therefore to put in place a system that could maximize the proportion of coffee available for export and restrict the amount of domestic consumption and marketing. With coffee production in the hands of smallholders, it was clear that control of distribution was a more expedient method of collecting the desired quantity of coffee than controlling production. In order to influence the course of coffee marketing, the government created in 1978 the Ethiopian Coffee Marketing Corporation (ECMC). The institution was given an initial mandate of buying coffee at prices geared to international market values,[36] but soon gained a much broader authority, which included: (1) the imposition of restrictions on the activities of licensed private traders; (2) the establishment of an official price structure at which the ECMC bought coffee and which formed a reference point for controlling prices in the private sector; and (3) the introduction of a system of physical controls on coffee movement to prevent the diversion of coffee to the domestic market and its export through illegal channels.[37]

The ECMC quickly controlled more than 80 percent of the officially handled exports. Despite substantial domestic demand, private traders were prohibited from selling coffee to domestic consumers. As was the case under the imperial regime, coffee auctions continued to take place, but prices at the so-called auctions were fixed by the Ministry of Coffee and Tea Development. Taxes on coffee were raised substantially and the forms in which they were collected multiplied. After surtax, transaction tax, export duty and coffee cess were paid, the total tax amount reached about half of the f.o.b. export value. In one carefully documented study of coffee marketing under the PMAC policies, it was demonstrated that the strict and high-tax control regime created strong incentives to smuggle coffee into the neighboring countries.[38]

With its strict system of controls, the government was able to achieve its goal of increasing Ethiopia's share of the international coffee market. The volume of exports increased from 47,058 metric tons in 1977 to 93,029 metric tons in 1983. The growth in exports resulted not from increased production but, by and large, from the draconian squeeze on domestic consumption. In fact, as production continued to experience a sharp decline, Harerge's contribution to Ethiopia's total coffee exports dropped dramatically, from 8,420 metric tons in 1976 to 2,260 tons in 1983, and reached its lowest level at 604 tons in 1985. In terms of the hectarage, coffee,

[36] The ECMC was established in 1975 under the National Coffee Board, mainly to further the government's objective of participating in coffee marketing, especially the export business. In 1978 the NCB became the Coffee and Tea Development and Marketing Authority, which subsequently was elevated to the status of the Ministry of Coffee and Tea Development (MCTD). The ECMC was then made a quasi-autonomous agency operating under MCTD.

[37] ULG Consultants, *Coffee Marketing, Processing, Transport and Storage Study*, vol. 3, unpublished report prepared for the Ministry of Coffee and Tea Development (1986), pp. 11.A.1–11.A.2.

[38] Stefan Dercon and Lulseged Ayalew, 'Smuggling and Supply Response: Coffee in Ethiopia,' *World Development*, 23 (1995), pp. 1798–9.

which occupied 6.6 percent of the area under cultivation in 1975, dropped to 4.4 in 1983.[39]

The decline in production was a consequence of farmers' reaction to physical controls, price fixing and prohibitive taxes. Recent studies on coffee production problems in Harerge have identified a variety of factors that affected coffee production. In a 1991 study, Girum Tamirayehu attributed coffee's production difficulties to:

> High [coffee] cess, purchase and export monopoly, high price paid for 'chat' [khat]..., coffee pests, the lower price paid to farmers, incompatible and unstable economic policies, food shortages leading to the replacement of coffee land by food crops and the competition of 'chat' (*Catha edulis*) which provides regular cash throughout the year have made substantial contributions to the decline of coffee production in Harerge.[40]

Since the late 1970s coffee production clearly faced a number of challenges, ranging from pest infestations to government intervention in its marketing and pricing. Farmers acknowledge that drought and disease (coffee berry disease or CBD) have discouraged them from planting coffee. At the time of my research in 1994 it was clear that farmers had replaced or were in the process of replacing their coffee trees with khat plants. Asked about why they had reduced the land under coffee cultivation, one informant replied with a series of rhetorical questions. He said:

> can you take one kilo of coffee to Harer [and sell it there]; can you take a kilo of coffee to Dire Dawa and get a fair price? Look! we were made thieves. We were reduced to stealers of our own property. Now what would you do if you were in my shoes? What would you do with a commodity that was made worthless by the government? Wouldn't you decide to abandon its cultivation and try to do something else for a living?[41]

The common theme in Tamirayehu's study and my informant's explanation is that the government's increasing effort to control coffee marketing drove farmers in Harerge to abandon its production. The price at which the ECMC purchased coffee was below production costs and discouraged farmers. The situation left farmers with only two options: protest the price-fixing policy or withdraw from producing the crop.[42] The informant's rhetorical questions suggest that coffee farmers made a conscious decision to increase their khat plantations rather than continue

[39] Eastern Ethiopia Zone Planning Office, *Bulletin of the Socio-Economic Conditions*, p. 8 and Appendix 9 and 10; Customs and Excise Authority, *Annual External Trade Statistics*, 1960–1991.

[40] Girum Tamirayehu, 'Coffee Production Problems and Cultural Practices in Harerge,' in Mesifin Ameha (ed.), *Proceedings of the First Workshop on Production Constraint Assessement of Coffee, Tea and Spices (CTS)* (Jimma, 1991), pp. 1–2.

[41] Interview: Mume Ahmed, Faldhaana, Kombolcha, June 8, 1994.

[42] According to Albert Hirschman, when deterioration occurs in organizations or members become dissatisfied, two alternative responses are available: exit – leaving the organization or switching to a competing product, and voice – agitating for change in policy, practice, or output from within. Albert O. Hirschman, *Exit, Voice and Loyalty: Responses to Decline in Firms, Organizations and States* (Cambridge, Mass., 1970), pp. 1–55.

to plant a cash crop that got diseased, fetched low prices and whose marketing was monopolized by the government. Farmers then preferred khat because government intervention in the industry was limited to collecting taxes and levying export duties. However, the socialist policies of state monopoly would soon extend to the production and exchange of khat too.

Traders versus the state

When the revolution began in 1974, there were fewer than 45 merchants who daily dispatched a total of 3,600 kg of khat to Djibouti. Despite the new government's avowed goal of restructuring agricultural production and exchange along socialist lines, the business of exporting khat remained in private hands for the first seven years of the revolution. From the government's perspective, khat was not so strategic a commodity as grain or coffee. In fact, any attempt at changing the existing pattern of exporting khat was seen as potentially disruptive of a business that generated so much revenue and foreign-exchange earnings. The government apparently saw no reason to increase its involvement in the marketing or pricing of khat.

The khat export business therefore remained dominated by a few prominent merchants. In a political environment inspired by the new government's professed commitment to standing up for 'the rights of the majority, not the privilege of the few,'[43] however, the merchants' grip on the khat export business did not go unchallenged. As early as 1975 there was a sudden surge in the number of applications for licenses to export khat. There was also an organized drive to make the khat business 'communal' and give it an 'Ethiopian character.'[44] The effort to end domination of the business by a few merchants was so strong that it was brought to the attention of the Chairman of the PMAC, Ethiopia's new head of state. As has been the tradition with the problems of khat marketing, the matter was referred to a commission appointed by the chairman of the ruling military council.

In its report, the commission acknowledged that the khat export system contradicted socialist principles and recommended the establishment of a new company comprised of small traders. This challenge to the oligarchic status of the khat export business did not bear fruit immediately. The private exporters protested strongly against the establishment of a competing organization, claiming that their enterprise provided work for 1,000 Ethiopians and earned close to 15 million birr in foreign currency for

[43] The applications are filled with mottoes of the leaders of the revolution. What the applicants were demanding was to be permitted to export khat since, as the revolutionaries were promising then, 'the period when a few elites dominated the country's means of production was a bygone era.'

[44] The phrase 'Ethiopian character' insinuated that the export business was dominated by Djiboutians, whom the organizers called 'French citizens.' HPA-GC.1, Six resident-traders of Dire Dawa to Samuel Teferra, Governor of Dire Dawa, Issa and Gurgura *awraja*, April 28, 1975.

the country. They also warned that creating a new company without the consent of the importing agency in Djibouti could result in a misunderstanding between the two companies and a disruption of the export business.[45] The private operation was thus permitted to continue as long as the tax revenue and foreign-currency earnings continued to flow into the treasury. The upshot of this whole episode was that, while the marketing and pricing of other agricultural products were brought under stricter government control, the khat business managed to remain relatively free.[46]

The tolerance of private khat exports was short-lived. With the post-imperial government's commitment to building 'a strong independent economy along socialist lines,' it was only a matter of time before all market-based operations were abolished. To achieve its goal of restructuring external trade in a way that would speed up the 'socialist transformation of commerce,' the government split the Ministry of Trade into the Ministry of Domestic Trade and a Ministry of Foreign Trade. The division meant differentiation of responsibility for procuring, processing and exporting commodities for domestic and foreign markets. According to proclamation 179/72, which established it, the Ministry of Foreign Trade was charged with the responsibility of 'putting the country's export trade on a socialist course by increasing the role of governmental and public organizations in the export trade sector, while streamlining, controlling and guiding the activities of private exporters.'[47]

Whereas the private export operation of khat was a target for nationalization, the need to control khat export was given an added impetus by the Ethio-Somali war of 1977–8. Considering the flourishing khat business as a source of employment for war victims and the families of officers who were disabled or died in the war, the *derg* government sought to gain control over the export sector and began organizing a government-run khat exporting company.[48] In March 1980 the Khat Exporters Cooperative of Ethiopia (KECE) was established under the direct supervision of the Minister of External Trade. According to the official statement that ordered its creation, the KECE's main stated goals were: (1) to impose order on the disorderly private export operations; (2) to bring the export business in line with socialist philosophy; and (3) to give the khat export trade an Ethiopian character and leadership. The charter of the KECE stipulated that the cooperative would be run by a board of directors, suggesting that the enterprise was a non-governmental private enterprise.[49]

45 HPA-GC.1, Testimonies of Hajji Hussein Roble and Ato Abduselam Hajji Mussa in the Minutes of the Committee that studied the conditions of khat trade, July 17, 1975.

46 HPA-GC.1, Minutes of the Committee that studied the conditions of khat trade, July 17, 1975.

47 MFT-CF.1, Eastern Ethiopia Ministry of Foreign Trade, Annual Report for the Years 1979–1990.

48 Interview: Captain Taddesse Debele, GM of the KECE (1994), Dire Dawa, February 23, 1994.

49 EHAZ-KF.1, Minutes of the meeting chaired by the Vice Minister of Foreign Trade convened to establish the KECE, March 20, 1980.

In reality, the government exercised power and influence over the cooperative. The general manager was appointed by the Minister of Foreign Trade, who also controlled one of the cooperative's bank accounts.[50] The cooperative was required to submit weekly performance reports to the regional representative of the Ministry of Foreign Trade.[51] Finally, the Minister of Foreign Trade reserved the authority of selecting and firing members of the cooperative. At the time of its creation, the cooperative had 263 members, including 128 from the previous organization.[52]

The KECE was also granted some important privileges often accorded to other semi-governmental institutions. Each of its members was exempted from customs, transaction and income taxes on all of the khat they exported and on 37.50 kilogram of khat they sold within the country.[53] PAs were assigned quotas to supply the cooperative with the khat it needed for export. As with its forerunners, the CECDD and the CHATHA, only the KECE was allowed to use Ethiopian Airlines to transport khat to Djibouti. The Ethiopian government negotiated and signed contracts on behalf of the cooperative with the government of Djibouti[54] by which the *Société Générale d'Importation de khat* (SOGIK), Djibouti's importing company, was bound daily to import 3, 670 kilograms of khat from KECE and not to buy from other exporters. The creation of the cooperative was clearly the first step along the road to creating a state monopoly of the khat export business.

Yet the *derg*'s own policy of using the khat business for political purposes had created an army of legally permitted khat traders who were not members of the cooperative. When Somalia invaded Ethiopia in 1977 law enforcement personnel, whose peace-time duties included safeguarding the Ethio-Djibouti railroad, were called into active duty for the more pressing task of fighting the war. The government sought to have local residents protect the railroad track in exchange for permission to supply khat to consumers along the line. The government also permitted more people to supply khat and other consumer necessities to people in the Ogaden, the majority of whom are Somali, to counter the message of Somali solidarity from Somalia. At the time it was thought that the threat of cutting off supplies would convince the Ogaden Somali to reject Somalia's propaganda and choose to stay within Ethiopia.[55]

After the war, the practice of supplying the Ogaden and the areas that had been under Somali occupation continued, with the aim of convincing refugees to return to their homes. The haphazard manner of supplying khat

[50] EHAZ-KF.1, Minister of Foreign Trade to the KECE, August 21, 1980.

[51] EHAZ-KF.1, Permanent Secretary of the Ministry of Foreign Trade to Eastern Harerge Branch Office, July 24, 1980.

[52] EHAZ-KF.1, Minister of Foreign Trade to the Eastern Ethiopia Branch Office, August 22, 1980.

[53] EHAZ-KF.1, Permanent Secretary of the Ministry of Foreign Trade to Eastern Harerge Branch Office, July 24, 1980.

[54] Eastern Ethiopia Ministry of Foreign Trade, *Report on the State of the Province's Foreign Trade* (Dire Dawa, 1989), pp. 4–6 (State of Foreign Trade hereafter).

[55] Interview: Beshir Shiek Abdi, Harer, March 24, 1994. In 1977, Beshir Shiek Abdi was Deputy Governor of Harerge province.

and other commodities was systematized and institutionalized. In November 1979 the government granted licenses to 581 Ethiopians and 1,579 Djiboutians to travel back and forth to Djibouti with fruits and vegetables worth 100 birr and import into Ethiopia various commodities valued at 100 birr duty-free.[56] These commuter-traders were permitted officially to conduct a limited import–export trade, but quickly became engaged in the business of unofficially exporting khat and smuggling in consumer commodities.[57]

The Ethio-Somali war thus created a favorable condition for expansion of unofficial trade in khat. The transfer of Ministry of Finance police and other law enforcement personnel to active duty weakened border control considerably. At the end of the war, only 10 checkpoints out of the 34 that had existed before the war remained operational, and they were so poorly managed that there was essentially no effective border-patrolling activity in the region.[58] Unlicensed traders infiltrated through the porous border while the licensed commuter-traders increasingly used the train to Djibouti to smuggle khat. A 1988 study estimated the number of such traders at over 4,000, only half of whom were licensed.[59]

Table 8.4 *Khat exports in volume & price in birr,* 1979–90

Year	Exported khat in kg	Total export value in birr	Unit export price in birr
1979/80	255,238	6,107.618	11.00
1980/81	2,399,392	26,393,317	11.00
1981/82	2,799,320	30,792,523	11.00
1982/83	2,816,143	30,977,573	11.00
1983/84	2,937,636	32,313,992	11.00
1984/85	2,768,441	30,452,856	11.00
1985/86	2,662,603	29,288,633	11.00
1986/87	2,575,229	28,327,523	11.00
1987/88	2,000,716	22,007,876	11.00
1988/89	2,358,497	27,793,120	11.784
1989/90	2,518,821	30,461,256	12.00
Total	26,391,936	294,916,287	11.17
Average	2,583,670	28,880,867	11.17

Sources: Abraham Gebremariam, Abebe Kebede and Yoseph Gebrehiwot, *Khat Exporters' Cooperative of Ethiopia: Progress and Prospects*, 1990; Eastern Ethiopia Ministry of Foreign Trade, *Annual Report for the years 1981–1991.*

56 MFT-CF.1, Eastern Ethiopia Ministry of Foreign Trade, Annual Report for the Years 1981–1990.

57 Interviews: Beshir Sheik Abdi, Deputy Governor of Harerge at the time, Harer, March 24, 1994; Captain Taddesse Debele, GM of the KECE, Dire Dawa, February 23, 1994.

58 Eastern Ethiopia Zone Planning Office, *A Comprehensive Report on Contraband Trade Activities and the 'Taiwan' Contraband Market* (Harer, 1988), pp. 4–6.

59 Minutes of the meeting of Harerge Province's high-level officials to discuss why contraband trade was on the increase, Harer, October 23, 1978, cited in Kidanemariam Gebreegziabher and Abay Asfaw, *A Study of the State of Illicit Trade in Harerge Region* (Amharic), Office of Central Planning for Eastern Ethiopia, Harer, August 1988, p. 25.

The KECE thus operated in a very competitive business environment that required speed and efficiency. During the first two years, the cooperative faced organizational problems ranging from inexperience, the lack of sufficient start-up capital and difficulties of procurement to sabotage by controllers, weighers and processors who were retained from the defunct private enterprises owing to their expertise but who were reportedly unhappy with the government-run business. Membership, as well as the amount of khat exported by the cooperative, did not increase substantially and no annual budget was adopted to facilitate the KECE's operations. The organization did not get off to a strong start as it joined the business of exporting khat from Harerge.[60]

Commencing in 1982, however, the KECE started to perform well. A new agreement was reached with SOGIK in which the amount of khat exported daily was raised from 3.7 tons to 8 tons. By 1990 membership in the cooperative had reached 810 and the number of employees working for the company grew to 1,500. Between 1980 and 1990 the KECE exported 26,391 tons and earned 295 million birr in foreign currency (Table 8.4), in contrast to 28,424 tons of coffee valued at 143 million birr and 69,649 tons of vegetables worth 37 million birr that were exported from Harerge during the same period. Measured primarily against the official economic indicators, by the end of the 1980s khat had become Harerge's most important cash crop.

The cooperative's success was largely a factor of government support at various levels of the export process. In the area of procurement, PAs and *awraja* administrators were made responsible for ensuring that the cooperative had purchased khat directly from farmers in quantities sufficient to comply with the terms of its contract. The cooperative maintained buying stations in Wobberra and Habro *awrajas*, and farmers were prohibited from selling khat to private traders before the cooperative's quota was met. Traveling with khat in a direction not leading to those buying stations was considered a felony.[61] Parallel to the coercive procurement methods, a public relations campaign was periodically conducted to persuade farmers about the virtues of selling khat to the KECE. Wole Chekol, the Minister of Foreign Trade in the 1980s, once told farmers' representatives: 'selling your khat to the KECE is an act of economic and political patriotism because your country benefits from the foreign exchange earnings and our economic adversaries will lose their leverage against our economy. Moreover, the KECE is a consistent customer, unlike the contraband dealers who buy khat wherever they find it inexpensively.'[62]

The government also protected the cooperative against competition from private traders, especially in curbing the ability of unofficial traders

[60] Abraham Gebremariam, Abebe Kebede and Yoseph Gebrehiwot, *Khat Exporters' Cooperative of Ethiopia: Progress and Prospects* (Amharic) (Dire Dawa, 1990).

[61] HPA-KF.1, Governor of Gara Mula'ata to Provincial Administration Office, September 28, 1988.

[62] Interview: Captain Taddesse Debele, GM of the KECE, Dire Dawa, February 23, 1994.

to smuggle khat and distort the existing price structure. Undoubtedly, the post-imperial regime had a better institutional capacity and reach than its predecessor in controlling smuggling. In addition to the traditional Customs Department and the Ministry of Finance police, the new regime had created an array of institutions in its effort to centralize the export economy. Ministries and organizations that, to various degrees and from various perspectives, had an interest in the khat industry included the Ministry of Foreign Trade, the PAs, the Urban Dwellers Associations (UDA), the various echelons of the Workers Party of Ethiopia (WPE), the Committees of Labor Inspection and Audit (CLIA) and the Provincial Administration. In some cases, units of the military (502nd Brigade, 9th and 11th Infantry Divisions) were deployed to combat unofficial trade activities.[63]

Despite the backing of this bureaucratic leviathan and its reliance on the government's coercive procurement strategy, the KECE did not achieve its goal of dominance in the khat export business. This was largely owing to its inability to compete with private traders in parallel markets. According to the contract between KECE and SOGIK, the latter paid a fixed price per kilogram of khat, thus hindering the KECE from paying producers a competitive price for high-quality khat. In 1985, for instance, while first-grade khat sold for 20 to 30 birr in Dire Dawa market, the cooperative was scrambling all over the khat production areas to find produce of similar quality for 2–3 birr. After transportation and processing costs were added to the stated purchasing price, the cooperative would end up spending more than the 11 birr per kilogram of khat that SOGIK had agreed to pay by the contract.[64] On the other hand, SOGIK could not increase the amount of khat it would accept from KECE because demand in Djibouti was saturated by supplies through unofficial channels. Unlicensed traders could sell first-grade khat for less than the cooperative's price, inducing more consumers to prefer smuggled khat. Their competitive advantage came from the fact that they incurred no expenses for tax and duties on both the Ethiopian and Djibouti sides.[65] The cooperative's reliance on state institutions to compensate for its inability to compete in the free market did not give it the advantage necessary to overcome the competition from unofficial traders.

The foregoing discussion clearly has shown that the competition between KECE and private traders was intense. Quantitative evidence on the activities of private traders who for the most part operated in parallel markets is hard to find. It is nevertheless possible to explain the cooperative's inability effectively to compete with private exporters by

[63] HPA-KF.2, Deputy Chairman of Harerge regional PA to all the 8 *awraja* PAs, January 2, 1986; HPA-KF.2, GM of KECE to Chairman of Dire Dawa *awraja* Workers Control Committee, February 18, 1986; HPA-KF.2, Commander of 11th Revolutionary Infantry to the Provincial Administration, August 13, 1986; EHAZ-KF.1, Eastern Harerge Shengo (parliament) to Harer Zuria *awraja* Administration, January 23, 1991.

[64] Kidanemariam and Abay, *Illicit Trade in Harerge Region*, p. 25.

closely examining official khat price trends and their relationship with figures showing supply conditions. Table 8.4 above shows the value of khat for the years 1981/2 to 1985/6 to be higher than the decade's average, whereas the volume of the exported product and price per unit remained more or less the same during those years. The amount of khat that was exported in 1990 dropped from the 1985/6 level by 56,500 kg, or by 2.19 percent, while price per unit increased by 7.6 percent over the 1985/6 level. The slight increase in unit prices from 11.8 birr to 12.00 birr, coupled with a market liberalization policy introduced in 1990 and improvements the company made in its procurement and marketing systems, had contributed significantly toward increasing the value of the exported khat. It is important to note the existence of inverse parallel trends in price–supply relationship pointing to structural constraints that prevented the company from being an effective competitor in the marketplace.

First, the company was unable to devise a mechanism that could stimulate farmer response to price incentives in order to obtain good-quality khat in sufficient quantities. Its failure was perhaps related to several factors, some of which were beyond the cooperative's control. Khat is a perennial crop and substantial lags (two to four years) exist before newly planted trees can be harvested. Therefore, price increases often do not result in more production. Likewise, price increases before harvest seasons could hardly stimulate production (and therefore supply) for the simple reason that farmers do not have the resources necessary to increase investment and productivity. The fact that supply remained virtually unchanged in the second half of the 1980s, especially after the unit price increases in 1988 and 1990, is evidence that farmers' response to price incentives was inelastic.

Second, the cooperative's inability to increase supply may have resulted in demand leveling off in Djibouti. Some surveys, though not scientific studies, estimate that approximately 15,000 kg of khat was consumed in the port city every day, nearly 40 percent of which was supplied through unofficial channels. The absence of competition from an organized khat supplier, rising prices in Djibouti (60 birr per kilogram in 1990) and the lack of a substitute for Ethiopian khat provided an ideal situation for the cooperative to command greater market share. With efficiency and competitiveness, the cooperative could have worked to expand the size of the market or gained ground on the parallel market suppliers. In Djibouti's case, however, demand might have leveled off as the result of weak supply response to rising demand and prices. In the late 1980s it was apparent that unofficial traders were gaining more share in the Djibouti market.

Third, because the KECE had to abide by its contract with the importing company, it was unable to out-compete private exporters for a larger market share in Djibouti. During high production months, when private traders who supplied Jijiga and the Ogaden tended to purchase untrimmed khat for as high as 4–5 birr, the cooperative was able to pay competitive prices and obtain good-quality khat for export. When khat became scarce in dry seasons, private traders could pay as high as 10 birr a kilogram and pass on the additional cost to consumers. Constrained by

Table 8.5 *KECE balance sheet,* 1987/8–89/90 (birr)

	1987/8	1988/9	1989/90
Total AwWoday khat	1,725,505	1,571,29	1,578,777
Processed khat per kilo of AwWoday khat	1.159	1.50	1.595
Exported khat	2,000,716	2,258,479	2,518,721
Unit price	11	11.784	12
Earned income	22,007,876	27,793,120	30,461,256
Expense per kilo khat	12.18	11.42	13.32
Total expenditure	21,016,648	17,944,230	21,029,308
Profit/loss	991,228	9,848,890	9,431,948

Source: Abraham, Abebe and Yoseph, *Khat Exporters' Cooperative*, p. 17.

its fixed-price agreement with SOGIK, the cooperative was unable to pay competitive prices. As we will see in the next chapter, the business environment was made even worse for the cooperative by smugglers who were willing to pay very high prices for the best-quality khat demanded by SOGIK.

The cooperative was gradually marginalized by the close commercial relations between smugglers and producers and by rising purchasing and processing costs. In 1990, expenses constituted 76.54 percent of the value of 1 AwWoday kilogram (often 1.5 kilogram in actual weight). The cost included expenditure on purchasing, taxes and transportation. Purchasing expenses per kilogram rose from 7.22 birr in 1985 to 9.06 in 1990, increasing on average by 10.63 percent each year. This increase was in part due to unforeseen expenses including the imposition of taxes in 1986 on KECE members and increases in stipends for purchasing agents and checkpoint guards as well as rising transportation costs. Of all the expenses per unit of khat, processing costs were the fastest growing.[66]

An analysis of the cooperative's balance sheet shows that the cost of preparing khat for export rose sharply between 1984 and 1990 while, as we have seen, the price SOGIK paid for 1 kilogram of khat remained stuck at 11 birr until it was increased to nearly 11.8 birr in 1988 (Table 8.5). Distributed over the years after the establishment of the cooperative in 1980, the price increase of 1988 amounted to less than 0.10 birr per year. Had the unit price increase not been made that year and in the subsequent years, the company would have declared bankruptcy and gone out of business.

If the structures of marketing and the cooperative's ways of operating were obstacles to its own success, how did it manage to remain afloat throughout the 1980s? The answer lies in the manner in which khat was purchased from producers and sold to consumers. As a general practice, khat was sold in bundles called *akara*, a bunch of khat that usually weighed

[65] Abraham, Abebe and Yoseph, *Khat Exporters' Cooperative* , p. 13.
[66] Ibid., p. 15.

1.5 kg or more at the farm gate. When the *akara* khat was processed, the weight obviously decreased but never fell below 1 kg. In 1987/8, the KECE drew on average only 1.15 kg from each *akara* purchased from farmers, which was lower compared with other years. Without the slight increase in selling price per kilogram of khat that year, the cooperative would have run a deficit of 3,761,598 birr. If the same analysis were done for the subsequent years, the company would have gone under as the result of the discrepancy between rising production costs and selling prices, which was never adjusted for inflation. Between 1988 and 1990, however, the cooperative was lucky enough to extract from each *akara* an average of 1.5 kg of exportable khat. It is therefore safe to say that the cooperative succeeded in remaining in the competitive khat export business by defrauding the producers.

Conclusion

The phenomenal expansion of Harerge's khat production in the 1980s was primarily the result of the land proclamation law of 1975, which permitted farmers to make their own decision about which crops to grow. Farmers also employed innovative farming strategies that helped to maximize productivity. It appears that Harerge's agriculture entered a qualitatively different phase when farmers began to purchase food and grow khat on land that otherwise would have been devoted totally to food crops. Khat's traditional rival cash crop, coffee, lost the competition for land, labor and capital investment when the *derg* government decided to monopolize its marketing while at first leaving khat to market forces.

The decision that farmers made to grow khat at the expense of food crops belies the belief long held by some scholars that farmers are totally unresponsive to market opportunities. The evidence shows that Harerge's farmers made their decision to expand khat after careful consideration of their farming resources, the market situation and the viability of investment in khat. The shift to khat cultivation was the result of a decision to cope with land scarcity and declining agricultural production. Harerge's farmers showed their willingness to abandon subsistence production in favor of cash crops that provided better returns on their investments. They sought to maximize their income and then invested it in nonagricultural economic activities. They made their choices based not solely on short-term market incentives, but also on political, environmental and family considerations.

Farmers make strategic decisions to safeguard the long-term welfare of their households rather than to obtain a temporary cash profit. In truth, farmers who practice low-input farming management have few resources with which to respond to price incentives. In a general way nonetheless it is possible to say Harerge's farmers shifted to the production of khat because it was more profitable than other crops. But such a blanket statement obfuscates the complexities of farmers' decision-making processes. The massive increase in khat production in the 1980s was less a matter of the simple cash profit that khat brought than its ability to provide money to buy more and better food. The availability of an alternative food source was therefore as crucial as the cash obtained from khat sales. It seems safe to say that Harerge's farmers did show some responsiveness to profit (not always cash) and shifted to cultivating the more profitable khat without violating the important principle of ensuring the availability of food.

Government policy was also an important factor in farmers' decision to increase the cultivation of khat at the expense of food crops. In all, increased government intervention in the production and exchange of agricultural products defeated the ultimate goal of more productivity and revenues and dampened farmers' enthusiasm for increasing productivity or production. Farmers who practice low-input agronomy are particularly apprehensive of government intervention. In areas where farmers have alternative marketing sources, it is difficult for a government to impose a command economy. Policy-makers may be well-advised to take notice of the findings of this research that farmers and traders have novel ways of getting around government plans they see as a threat to their economic well-being.

Nine

Khat & Contraband

Profits, professions & parallel markets

> Allied to the chat [khat] market, and financed by it, is a flourishing trade in contraband goods brought in from Djibouti, with the result that Hararghe forms a kind of free trade zone within a socialist state.
>
> — Christopher Clapham, 1988

Throughout its recent history, Harerge has maintained stronger economic links with the coast than with the hinterland. In the 1980s this economic relationship was characterized by the hidden export of primary products and the import of consumer commodities. The value of this activity is somewhat difficult to calculate, but several studies have estimated smuggling in the early 1990s to be worth as much as 1.05 billion birr a year. Such a high figure is not completely implausible, considering the ubiquitous presence of contraband goods all over Ethiopia, from Dire Dawa's 'taiwan' market to Addis Ababa's 'marcato' and to small town kiosks in the western provinces. No comprehensive scholarly work has yet been published on this phenomenon in Harerge, although there are passing remarks in a few works dealing with broader political and economic issues. Perhaps the most succinct description of the relationship between khat and contraband trade was given by Christopher Clapham, a long-time observer of Ethiopian politics. In his seminal study of the Ethiopian revolution he commented that

> Hararghe constitutes, to an extent impossible to appreciate without personal observation, an economy based on chat [khat]. ... Farmers can make about 3,000 birr a year from the crop, putting them well above the standard of living of highland peasants in other parts of the country. Allied to the chat [khat] market, and financed by it, is a flourishing trade in contraband goods brought in from Djibouti, with the result that Hararghe forms a kind of free trade zone within a socialist state. The economic frontier of Ethiopia stands in effect on the Hararghe/Shoa border at Awash, and the region as a whole stands in the sharpest

contrast to the problems of agricultural production and marketing in other parts of the country.[1]

As Clapham rightly noted, contraband activities provided a solution, albeit temporarily, to the problems of agricultural production and marketing generated by interventionist government policies and offered an autonomous employment opportunity in the context of limited entrpreneurial ventures.

In the 1980s contraband trade flourished in Harerge. The market stimulated khat production and trade, and most of Harerge's farming communities consequently became awash with khat-generated money. Smuggling provided employment for a considerable number of urban residents. By the end of the decade, contraband trade had expanded to the extent of undermining the fiscal, judicial and political credibility of many state institutions in Harerge. Constituting half the value of all the products that were smuggled out of the province, khat played a very significant role in making the cross-border trade possible and in improving the standard of living for those who took part in the phenomenon of contraband trade.[2]

The national & regional economic context

In Harerge the realm of economic activities that lie beyond a government's taxing, regulating and measuring capacity is referred to as contraband. In common parlance, the term connotes illegality, although the goods that are involved in the commercial transactions are perfectly legal. The interest of authorities was to collect unpaid dues and taxes and enforce pertinent regulations, rather than to stop output or exchange. Thus, illegality stems from the contraband traders' conscious evasion of taxes and regulations. In a way, the contraband trade is an alternative form of work for those whose livelihood was threatened by government intervention.

The boundary between legal and illegal commodity exchanges in Ethiopia was delineated by the Customs Proclamation of 1947, which defined contraband trade as 'any importation or exportation of goods with the intent to defraud the revenue or to evade any prohibition of restrictions on, or regulation as to the importation and exportation.'[3] Further, 'all smuggled goods, all goods imported in contravention of any prohibition, restriction or regulation, all goods that are dutiable found in any aircraft or other means of conveyance which have not been declared to the customs …'[4] were categorized as contraband. From a strictly legal

[1] Clapham, *Transformation and Continuity*, p. 186.

[2] Getachew Eshete, 'Contraband Trade in Ethiopia: Major Highlights,' Paper Presented at the Seminar on Contraband Trade, Addis Ababa, July 1995, p. 8.

[3] Customs Proclamation No. 145/47. The Criminal Code goes further, defining what constitutes contraband trade and listing the penalties imposed on any person found guilty of smuggling.

[4] *The Consolidated Laws of Ethiopia*, vol. 1, as quoted in *Meskerem* (monthly magazine), 16 (1984), p. 40; *Negarit Gazetta* 10 (1941).

standpoint, even the cultural exchanges that had gone on for centuries between the Harerge highlands and the coastal regions came under the broad rubric of contraband, although they had been perfectly legal before the proclamation. Ironic as it may seem, it is the creation of Customs as a government institution responsible for regulating and collecting duties on export trade that gave rise to contraband trade. In a legal sense, therefore, it is logical to state that contraband trade in Harerge came into existence in the early 1940s.[5]

In the 1950s and 1960s the imperial regime attempted to curb contraband trade by deploying a system of controls characterized by thorough searches of traders at numerous checkpoints on the major routes leading to and at border crossing points. Licensed private traders imported consumer goods such as cigarettes, cosmetics, textiles, wrist watches, shoes and so forth, but their capacity was too limited to satisfy the region's demand, much less that of the neighboring provinces. In the two decades preceding the downfall of the imperial regime, contraband trade expanded as traders increased their effort to evade redundant checkpoints and excessive taxes and in response to the rising demand for consumer commodities that were not available in the country in sufficient quantities.

In the 1980s contraband trade became enormous in volume and value, different in its causes and sophisticated in its operations. The most important factor accounting for its growth was the wide disparity between supply and demand of basic consumer goods such as clothing and food products created by changes in Ethiopian society. In the decades since the expulsion of the Italians, Ethiopian society had undergone significant demographic and socioeconomic changes. The growth of modern state institutions and the expansion of private firms had produced a sizeable urban population and had monetized the economy, thereby profoundly altering consumption patterns. The demand for commodities thus generated was not met by domestic manufacturing, and the country's import-substitution policy did not favor the importation of the commodities in demand to make up for supply shortfalls. The gap between demand and supply was especially wide in the lowlands and border areas of Harerge. In the early 1980s, government and public institutions were responsible for distributing consumer necessities in four lowland *awrajas* of Harerge. According to figures generated by the regional Office of Planning, the amount of goods government institutions were able to distribute was rather low.[6]

The figures in Table 9.1 show that there was a significant gap between demand and supply. The estimated total demand for consumer goods for 1987/8 was 122 million birr for the four *awrajas* while the

[5] Neville Williams, *Contraband Cargoes: Seven Centuries of Smuggling* (London, 1961), pp. 1–28.

[6] These figures are not accurate measurements, but indicators of a trend in the relationship between demand and supply in Harerge's lowlands. Abebe Kebede, Abebe Mekonnen, Mesfin Shiferaw, Samia Yusuf and Teferi Tilahun, *The Causes and Characteristics of Illicit Trade and Future Trends in Eastern Ethiopia*, Dire Dawa, 1989, pp. 59–60.

Table 9.1 *A comparison of potential demand & actual supply of commodities for four lowland* awrajas *of Harerge in '000 birr*

Awraja		1987/88				1988/9			
		(1)	(2)	(3)	(4)	(1)	(2)	(3)	(4)
Kebri	demand	14648	909	4424	19981	15824	922	4552	21297
Dehar	supply	531	1008	135	1674	486	369	59	914
	%	3.6	110.1	3.0	8.4	3.1	40.1	1.3	4.3
Warder	demand	40047	2410	11039	53497	43032	2489	11404	56923
	supply	896	420	89	1406	259	366	21	645
	%	2.0	17.0	-	2.0	-	14.0	-	1.0
Gode	demand	14345	822	4124	19290	15361	847	4219	20426
	supply	658	755	135	1547	506	565	57	1130
	%	4.0	91.0	3.0	8.0	3.0	66.0	1.0	5.0
Kelafo	demand	14632	836	4055	19523	16679	866	4174	21718
	supply	556	356	96	1008	415	80	63	559
	%	3.0	41.0	2.0	5.0	2.0	9.0	1.0	2.0
Total	demand	83671	4978	23642	112200	90894	5124	24346	120365
	supply	2641	2358	455	5634	1693	1380	203	3248
	%	3.0	50.0	2.0	5.0	1.0	2.6	-	2.0

Source: Abebe *et al.*, *Causes and Characteristics of Illicit Trade*, p. 59. (1) *Food products* comprised mainly sugar, flour, pasta/macaroni, rice and cooking oil. (2) *Other commodities* included mainly soap, detergent, matches and battery cells; (3) *Textiles* were mainly *abujedid*, poplin, garments. (4) *Total demand* for 1, 2 and 3.

amount of goods distributed by government enterprises was worth 5.6 million birr, or just 5 percent of the projected demand. Such a wide gap shows not just the inadequacy of the system of distribution but also the enormity of the problem even if a serious effort were to be made to meet the need for consumer commodities in the lowlands. Corroborating the trend shown by the figures is a study conducted in 1982 under the auspices of the Minister of the Interior, which discovered that Harerge's lowland *awrajas* and border areas hardly received consumer necessities from the Ethiopian side. The study confirmed that consumer commodities were in short supply in Harerge's lowlands owing not only to the inability of domestic manufacturers to produce in quantities commensurate with demand, but also to the distributional inefficiencies of cumbersome government organizations. The Ethiopian Domestic Distribution Corporation (EDDC), the government agency that distributed consumer goods, maintained a few distribution outlets in the lowland towns, but bypassed most border communities. Private traders attempted to supply these remote towns, but their prices often remained beyond the reach of the residents there. In order to obtain the goods they needed, the lowland residents forged close commercial ties with the communities in Somalia and Djibouti with whom they were ethnically affiliated.[7]

Shortages of consumer necessities were also evident throughout Ethiopia. Conventionally, when domestic production fails to satisfy the existing demand, imported goods would make up for the shortfalls. In the 1980s, however, the Ethiopian government pursued a monetary policy that denied private exporters access to the country's foreign currency resources. The only way private traders were permitted legally to import consumer commodities was an arrangement called the franco valuta (FV). The system allowed for private imports that did not involve remission of hard currencies from the National Bank of Ethiopia.[8]

The franco valuta arrangement was not without restrictions. Commodities such as electronic appliances, woolen garments, perfumes, cigarettes and so forth were excluded from the list of importable goods, classified as luxury products, even though no hard currency from Ethiopia was involved in importing them. Moreover, even when official imports were allowed, private traders were subjected to onerous taxes and payments. In 1982, in addition to a 219 percent surtax and 5 percent import levy, the National Bank of Ethiopia, the Maritime and Transit Authority, and the Port Authority charged service and storage fees. When all was added up, a private trader would have paid as much as 269 percent in taxes and service fees before any consumer goods had cleared through Customs.[9]

Prices on goods imported in this way soared to unaffordable levels when they reached the market. The ordinary Ethiopian consumer preferred the less expensive contraband goods, thereby creating high demand for such smuggled commodities as electronic equipment, fashion accessories, firearms, clothing, canned and packaged food and household utensils. It is hard to know the accurate proportion of the goods that were actually imported, but it is possible to project estimates from the quantities of seized goods. In 1987, for instance, the Ethiopian Customs Authority seized goods worth 37.5 million birr, of which textiles, clothing (new and second-hand) and footwear constituted a remarkable 68.6 percent, foodstuff and drinks 4.0 percent, and electronic appliances, cosmetic products, cigarettes and matches together almost 15 percent.[10] By the Customs Department's estimation, the proportion of the goods seized was somewhere between 1 and 10 percent of the goods actually smuggled. This means that in 1987 goods worth between 375 million

[7] Ministry of Interior, 'The demand for and distributional difficulties of consumer commodities in border *awrajas* and recommended solutions,' A report to the Minister of Interior (in Amharic), Addis Ababa, 1984.

[8] Dejene Aredo and A. Abduraheman, 'A Study of the Contraband Trade in Eastern Ethiopia,' unpublished document, Department of Economics, Addis Ababa University, n.d., p. 4.

[9] Hagos Gebre, 'Contraband Trade in Ethiopia' (Amharic), Prepared for the Office of the National Committee for Central Planning (Addis Ababa, 1988), pp. 7–8.

[10] Dejene and Abduraheman, 'Study of the Contraband Trade,' p. 3.

birr and 3.8 billion birr were smuggled into Ethiopia. An average of these apparently extreme figures seems to correspond to the 1.05 billion birr estimated quantity of goods smuggled into the neighboring countries of Somalia and Djibouti.[11]

Statistics indicate that at least 44.6 percent of the total seizures came in Harerge region. At a 1 percent seizure rate, consumer goods that entered Harerge would be worth 167.3 million birr while at 10 percent, the value rose to 1.7 billion birr.[12] These figures obviously do not represent the precise value of smuggled goods, but they are not arbitrary either. As projections based on seizure statistics, if anything, they show trends. It is obvious that the gap between the lowest and highest projections is large, but, even at the lowest projection of 167.3 million birr, the contraband trade was a very significant business, unmatched by the official export trade from Harerge, which peaked at 75 million birr in 1983.[13]

The foregoing examination of the national and regional economic contexts was meant to outline the circumstances that led to an unprecedented expansion of contraband trade activities in the 1980s. It has been shown that the economic policies of the post-imperial regime were so contrary to free-market exchanges and transactions that they created supply shortages in consumer commodities. Traders took advantage of the situation to supply the residents of Harerge highlands and the rest of Ethiopia with the goods in demand through unofficial imports.

Khat & its role in parallel markets

Khat plays a critical part in Harerge's parallel markets in the sense that its unofficial exportation finances the unofficial importation of commodities. According to the seizure statistics of 1987 cited above, the stimulant accounted for 50 percent of the primary products smuggled out of Ethiopia. In Harerge alone, however, most of the local sources consulted show that live animals (cattle, sheep and goats as well as camels) constituted the largest proportion (66.9%) of the smuggled-out products, followed by khat (28.5%) and coffee (4%). Based on the estimated value of smuggled products given above, the percentage value for khat amounts to 300 million birr in real currency.[14]

Except for khat, all other products were smuggled into the neighboring countries and then reexported. Live animals, especially sheep and goats, were shipped to the oil-rich countries of the Arabian Peninsula

[11] Office of the Planning Committee for the Western Harerge Administrative Area, *Guidelines to Control Contraband Trade in Livestock in Western Harerge Administrative Area* (Asba Tafari, 1989), pp. 6–9 [OPCWH, *Contraband Trade in Western Harerge* hereafter].

[12] Getachew, 'Contraband Trade in Ethiopia,' p. 5.

[13] MFT-CF.1, 'Annual Report of the Eastern Ethiopia Branch Office for 1986–87.'

[14] OCPWH, *Contraband Trade in Western Harerge,* pp. 6–9; Abebe *et al.*, *Illicit Trade in Eastern Ethiopia,* pp. 67–75.

and sold there for hard currency. Coffee was purchased in Harerge and smuggled into Djibouti, making the latter a coffee exporter even though it grew none. The proceeds from the sales of these commodities were used to purchase such consumer goods as textiles, electronic appliances, footwear, wrist watches, battery cells, fragrances, incense sticks, cigarettes and so forth for smuggling back into Ethiopia.[15] Buyers who ultimately smuggled out live animals or coffee paid much higher prices to producers than the respective government institutions that had the legal priority to purchase those commodities. In 1983, for example, an ox was sold for 11,000 Somali shillings, or 1,100 birr, in Somalia, almost ten times higher than the price the Ethiopian Meat Corporation offered.[16] Likewise, the Ethiopian Coffee Marketing Corporation offered coffee producers on average 41 percent of the f.o.b. value.[17] The depressed values offered by the government institutions were reasons enough for producers to prefer having unofficial traders as their favored customers.

Khat smuggling was particularly attractive because of the enormous differential between producer and consumer prices. Throughout the 1980s, as we have seen, the highest price the KECE was able to pay for one kilogram of khat was 7.50 birr. Even this amount was reached in 1990, only after SOGIK raised the price it paid to 12 birr/kg. For most of the 1980s the KECE paid producers an average of 3 birr/kg of khat while first-grade (*urata* and *quda*) product was sold in Djibouti for as high as 120 birr. The third-grade (*qartii*) khat which the KECE exported was sold in Djibouti by SOGIK for an average 60 birr/kg. At an average producer price of 5 birr and consumer price of 80 birr, khat prices rose by a staggering 1,500 percent between the farm gate and the Djibouti market.[18]

Traders preferred khat smuggling primarily for the profit they made in a single run across the border. It also served yet another important purpose of enabling traders to earn their own 'foreign currency,' which they used to finance the imports of consumer goods in short supply at home. The practice of earning one's own foreign currency was widespread throughout the late 1970s and 1980s, but it became especially important after May 1988, when the use of the franco valuta arrangement was banned, allegedly owing to too many luxury commodities entering the country and the government's conclusion that the system benefitted smugglers more than ordinary consumers.[19]

In the 1980s, therefore, khat smuggling was more than an attempt to evade high taxes, excessive regulations and bureaucratic corruption. It

[15] HPA-KF.2, Chief of Security for Harerge to the Office of the Chief Regional Administrator, October 9, 1982.

[16] HPA-KF.2, Chief Administrator of Garamulata to Harerge region's Administration Office, May 18, 1983.

[17] Hagos, 'Contraband Trade in Ethiopia,' p. 6.

[18] Kidanemariam and Abay, *Illicit Trade in Harerge Region*, pp. 23–5.

[19] Dejene and Abduraheman, 'Study of Contraband Trade,' p. 4; Hagos, 'Contraband Trade in Ethiopia,' pp. 7–8.

was also a means to get access to hard currencies with which to import much-needed consumer commodities: the volume of the goods that were smuggled out was determined by the value of the goods that were smuggled in. The ubiquity of unofficially imported consumer goods in Harerge and the rest of Ethiopia suggests that a significant amount of capital was involved, but, since smuggling took place outside the official sector, these goods do not appear in the national statistics. It is difficult to determine precisely how much khat was exported and how much foreign currency was obtained therefrom.

It is possible, however, to collate suggestive statistics and obtain a reasonable estimate of the amount of khat smuggled into Djibouti in the late 1980s and the value of goods imported in exchange. According to a 1987 survey, 5,350 kg of khat was smuggled daily into Djibouti by train passengers, licensed commuter-traders and contraband traders known locally as *kotébélé*.[20] This figure is modest and entirely credible for three reasons. First, on June 6, 1988, the regional administration mobilized government as well as public institutions and conducted a surprise search along the railway line and along the major contraband routes. On that day, 9,710 kg of khat intended to be smuggled into Djibouti alone was confiscated.[21] Second, the survey figure is consistent with the 15,000 kg of khat that was estimated to be consumed in Djibouti daily throughout the 1980s. The KECE supplied 8,000–9,000 kg, and the rest was supplied through smuggling. Third, the estimate is fairly close to the World Bank estimate of 8,000 kg per day that was smuggled out of Ethiopia.[22] Local sources indicate that 10,000 kg of khat was cleared through the Jijiga customs, out of which 4,000 kg was smuggled daily into Somalia through Harshin, Togochale and Tafari Ber and another 4,000 kg was clandestinely taken to Djibouti through Durdur, Dikil and Dewalé.[23] The total is exactly the same as the World Bank estimate, which, unlike the local sources, had taken into account surveys done on the situation in the receiving countries.

It is important to note that local sources and a survey of World Bank experts agree that a total of 8,000 kg of khat was daily smuggled into Djibouti and Somalia. On the basis of this figure, 2,920,000 kg of khat was smuggled out of Ethiopia in 1987/8 and perhaps every year throughout the 1980s. At an average rate of 60 birr per kg, which is the lowest Djibouti price paid for a third-grade khat, the value of smuggled khat was 175,200,000 birr. This amount, it must be noted, was calculated at the lowest price per unit and must be seen as a fairly accurate estimate of the value of khat smuggled out of Ethiopia.[24]

[20] The name refers to unlicensed traders who travel back and forth to Djibouti climbing on train tops to get free passage. They have been the main conduits of contraband since probably the late 1950s, when the volume of the trade increased considerably.

[21] Abebe *et al.*, *Illicit Trade in Eastern Ethiopia*, p. 67.

[22] World Bank, *Ethiopia: Export Action Program*, (Washington, 1987), p. 47.

[23] Abebe *et al.*, *Illicit Trade in Eastern Ethiopia*, pp. 67, 81.

[24] Ibid., p. 68.

This sum is the birr equivalent of either Djibouti francs or Somalia shillings with which the traders were paid for their khat. Occasionally, the traders exchanged their francs or shillings for US dollars, but in most cases they purchased consumer goods to smuggle back to Ethiopia. The 175 million birr figure reflects the Djibouti market price for khat, not the actual value of the consumer commodities sold in Ethiopia. Even at a modest profit rate of 20 percent, the goods that were smuggled into Ethiopia were worth 210 million birr, or US$ 42 million at the black market rate of 5 birr that prevailed during much of the 1980s.[25]

These are astonishing figures. Perhaps many will suspect their accuracy, given the size of the official national economy. Although figures generated in different situations were cross-checked and show remarkable consistency, the final result must be viewed within two important contexts. First, during the 1980s, the parallel, rather than the official, market was growing in Harerge. The contraband trade was the response of Harerge's population to widespread unemployment and under-employment and the falling value of real wages. Although some of the capital that circulated in the parallel market was owned by a few wealthy merchants, a considerable number of ordinary Ethiopians were involved in smuggling. Second, the parallel market was self-sustaining and capable of generating high rates of accumulation and of existing as a semi-autonomous economic system. In other words, the size of the unofficial economy should not be taken as the benchmark to determine the size of the parallel markets since the latter could outgrow the former, whose growth was constrained by official policies.

Smuggling is a risky business with the possibility of high rates of return. Profit came to smugglers not only from the sale of khat and other primary commodities but also from the sale of consumer commodities back in Ethiopia. Although the government suffered loss of tax revenue and foreign currency, the contraband trade helped to alleviate supply shortages of consumer commodities. An estimated 1,253 *kotébélé* and 600 commuter-traders directly participated in smuggling, and a considerable number of people found employment in the industry as brokers, porters, khat processors and drivers. In indirect ways, local governments also benefitted from the contributions the contraband traders made toward various development projects. In 1985/6 and 1986/7, the city of Dire Dawa collected 3.6 million birr in contributions from traders while the transportation and hotel industries in the city expanded in response to rising demand.[26]

[25] At the official exchange rate of 2.07 birr to the dollar, the value of goods smuggled to Ethiopia in 1987/8 would be $101 million.

[26] The smuggled goods were sold in Dire Dawa in the 'taiwan' open market and in private stores. The owners paid taxes and made several contributions toward building schools, government offices and assisting drought victims. Abebe *et al.*, *Illicit Trade in Eastern Ethiopia*, p. 77.

The infrastructure of contraband trade

Harerge's parallel markets generated considerable employment as the result of the dynamic circulation of capital and commodities. Rural households in Harerge became enmeshed within a parallel market economy in which they contributed twice to capital accumulation: first, as producers of khat leaves, but in a process where much of the value was appropriated at stages closer to the point of consumption; and, second, as conspicuous consumers of contraband goods used to repatriate profits back into the country. Although farmers were not the foremost beneficiaries of the khat prosperity of the 1980s, they had improved their standard of living considerably.

Those who chose the risky, but highly lucrative, path of contraband trade in Harerge were not necessarily, as one might suspect, traders of dubious reputation. The first category of participants was comprised of established merchants, some of whom were of considerable means. They were not involved in physically smuggling goods, but they provided the initial capital to purchase khat and extended advances to people who did the work of physically delivering the product. Unlike the smuggling of live animals, which was a well-organized operation, the khat business was not tightly controlled by big merchants. Since most khat smuggling was done by small operators, its benefits were somewhat evenly distributed among them.

The khat smugglers formed another principal category, and they came from diverse socioeconomic and ethnic backgrounds. Some were engaged in the business full-time, delivering khat to destinations in the neighboring countries and transporting back consumer goods. Comprising this group were urban school dropouts, and young men who had left home to escape National Military Service. Most of them came to Dire Dawa from other provinces and joined the only occupation that required no qualifications or licensing. The other full-timers were members of the Issa ethnic group who inhabited the area stretching from Afetessa in Dire Dawa, along the railroad to Dewalé at the Ethio-Djibouti border. The Issa have always participated in the trade across their land, mainly as transportation providers, and were considered 'professionals' and known as *kotébêlé*. In 1978 the number of Issa who regularly participated in the contraband trade was estimated to be two thousand.[27]

Then there were people who obtained permits to conduct limited import–export trade during and after the Ethio-Somali war of 1977. In 1978 approximately 600 disabled veterans of the war were permitted to travel to Djibouti with fruit and vegetables worth 100 birr each and, in exchange, import into Ethiopia various commodities of the same value duty-free.[28] About 1,200 Djibouti nationals were also given *laissez-passers*

[27] Tesfamichael Gebremedhin, *The KECE's Chief Inspector's Report on the State of Illicit Khat Trade along the Djibouti Railway Line* (Amharic) (Dire Dawa, 1986).

[28] Minutes of the meeting of the Harerge Administrative region's high-level officials to discuss why contraband trade was on the increase, Harer, October 24, 1978, in Kidanemariam and Abay, *Illicit Trade in Harerge Region*, p. 25.

to travel to Ethiopia and import the same amount of fruit and vegetables. A total of 1,800 persons were authorized to commute between Dire Dawa and Djibouti to conduct officially prescribed amounts of trade between the two countries. Over the course of time, these commuter-traders found khat to be more profitable and they increasingly engaged in smuggling it. In 1980 in order to curtail their involvement in parallel markets, the government required the commuter-traders to obtain licenses in order to export farm products, fruits and vegetables, spices and processed food products worth 500 birr and to import consumer goods with the exclusion of cosmetics, jewelry, wrist watches, liquor, woolen textiles, tape recorders, radios, cameras and various kinds of incense.[29]

The excluded merchandise comprised those in high demand. Since there were no facilities in place to weigh or measure, the commuter-traders often imported or exported goods in far greater amounts than the officially permitted level. Data indicate that the commuter-traders were deeply involved in the contraband trade. Between 1983 and 1985 the commuter-traders officially exported between 2.5 and 3 million kilograms of fruit and vegetables. They traveled to Djibouti more than twice a week, but, even so, the amount of fruit and vegetables each commuter-trader took on a single trip on average was only 13.5 kg, 76.5 kg less than what an individual trader was legally permitted to export.[30] It is unlikely that the traders were taking with them amounts that were not sufficient to pay even for their transportation, much less to sustain their livelihood. In 1981 the government repealed the law that permitted limited import–export by commuter-traders, but, since the mechanism of enforcing the regulation was inadequate, the change did not do much to stop the commuters from taking part in the unofficial trade.[31]

The third group of smugglers were current and former employees of the Ethio-Djibouti railway company, the Ministry of Finance police and the Customs Department. Except for the former employees of these institutions, those who participated in smuggling included train engineers, station masters, passenger inspectors and the railroad company's security personnel. Members of the police and Customs workers rarely participated in actually smuggling goods but played a vital role in allowing smuggling in return for graft.[32]

Most of the smuggled khat was processed in Dire Dawa, although

[29] Ministry of Foreign Trade, 'Guidelines for Regulation of Import–Export Trade' (Dire Dawa, 1988).

[30] That the traders were not actually acting within the legal limit can be demonstrated by the following calculation: (a) the amount of fruits and vegetables exported between 1982 and 1984 was 2,531,899 kg; (b) the number of commuter-traders was 1,800; (c) the average number of trips to Djibouti per week was 2; 2 trips x 52 weeks = 104 trips. a ÷ b x c = 13.5. Kidanemariam and Abay, *Illicit Trade in Hararge Region*, p. 17.

[31] EEZPO, *'Taiwan' Contraband Market*, *inter alia*; Kidanemariam and Abay, *Illicit Trade in Hararge Region*, p. 18.

[32] Abebe *et al.*, *Illicit Trade in Eastern Ethiopia*, pp. 82–3; Getachew, 'Contraband Trade in Ethiopia,' p. 4; Tesfamichael, *State of Illicit Khat Trade*, pp. 10–11.

from the mid-1980s on Shinille and Melka Jebdu, two villages that grew to being major commercial hubs, also served as processing sites. Melka Jebdu is located northwest of Dire Dawa, and Shinille was a small train station on the railway line to Djibouti.[33] In the 1980s the two villages grew to town status as more people migrated there, built residences and established new businesses. The towns were used by smugglers as staging centers and transfer points for incoming and outgoing contraband goods. In these two towns and in Dire Dawa, smugglers processed khat in private homes, in coffee and tea parlors and in hotel rooms.[34]

To transport the processed khat to the border, smugglers used the Djibouti-bound train most of their way. Some khat originated in Dire Dawa, where a few of the smugglers jumped over the fence surrounding the train station and boarded locomotives, while a few others climbed on the train as soon as it left the station and before it picked up speed. Most of the khat processed in Dire Dawa and Melka Jebdu was gathered at Shinille after being transported there by pick-up trucks, pack animals and human porters. In accordance with its schedule, the train stopped at Shinille briefly and smugglers got on board with their khat.[35] From Shinille the smugglers traveled to Laserat, 150 km away, where most of the khat was unloaded and taken 50 to 65 km to Durdur, on the Djibouti border, or 20 to 30 km to the Somali border. At the borders khat was transferred to clients, who rushed the leaves to Djibouti by speedy trucks. The remaining khat was taken to Ayisha, a small station, where the train stopped for 10 minutes, and Dewalé, a major immigration and customs center, where searches were supposedly thorough and complete. The *kotébélé* disembarked from the train before it entered Ayisha and crossed the border on foot. Others, mainly the commuter-traders with *laissez-passers* and licences, and the railway company workers, the customs and police personnel, managed to get through the search process at Dewalé. Some khat, of course, was confiscated but most was successfully smuggled. The experienced smugglers were so confident about clearing customs that one bewildered KECE inspector said: 'the smugglers do not seem to be aware that smuggling contraband goods is against the law. They do not really seem to mind that their action is a punishable offense.'[36]

In the 1980s the use of the railway for contraband trade purposes increased considerably. Passenger and freight cars were frequently used for smuggling khat. The traders opened up and stored their khat in the ceilings and the walls of locomotives to hide it from authorities. Officials

[33] Ministry of Foreign Trade, Branch Office for Eastern Ethiopia, 'Report on the State of the Province's Foreign Trade,' Dire Dawa 1989, pp. 6, 14 [MFTEE, 'The State of Foreign Trade,' hereafter]; Ministry of Foreign Trade Branch Office for Eastern Ethiopia, 'A Report on Ways to Control Contraband Trade at the Train Station and Safeguard the Train,' February 16, 1989, pp. 4–5 [MFTEE, 'Control Contraband Trade' hereafter].

[34] MFTEE, 'Control Contraband Trade,' p. 4.

[35] Tesfamichael, *State of Illicit Khat Trade,* pp. 1–6.

[36] MFTEE, 'Control Contraband Trade,' pp. 10–11.

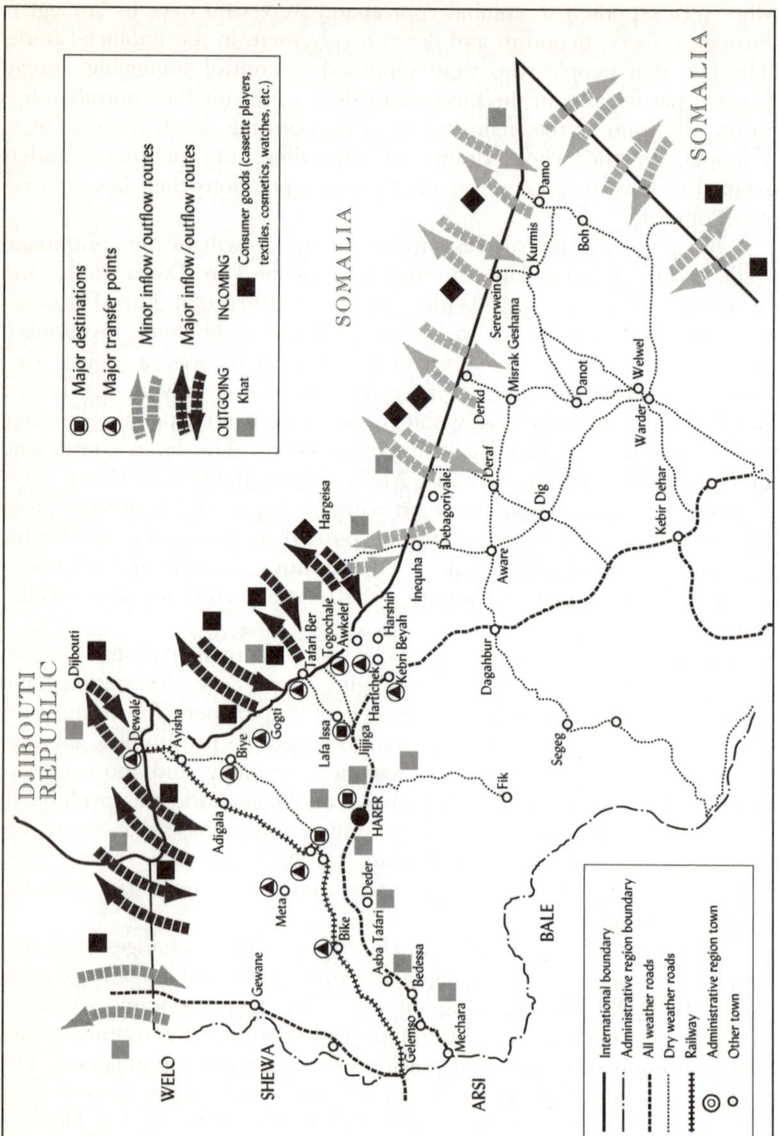

Map 9.1 *Routes of smuggling*, 1980s

who were supposed to combat contraband were won over by smugglers through bribery, nepotism and direct involvement in the unofficial trade. The fact that people who were supposed to control smuggling instead became participants in the business made it easier for the contrabandists to use the train as the main means of transporting goods in and out of Ethiopia. In the whole drama of smuggling, the unofficial traders seemed to have more say over the railroad operations than law enforcement officials.[37]

Corruption was in part responsible for the growth of the contraband trade because it led to lapses in inspection at the Dire Dawa station and at Dewalé, the last stop before the train proceeded into Djibouti's territory. As for the areas in between, officials deliberately overlooked contraband activities. After October 1978 there was a policy that unofficially imported goods were to be confiscated only at the border point. Once the goods were within Ethiopian territory, people were free to buy and sell the unofficially imported goods. The main contraband markets at Dire Dawa ('taiwan'), Melka Jebdu, Jijjiga and Harer, were outdoors and transactions went on with no apparent interference from law enforcement.[38] The major customs post where goods were thoroughly searched and confiscated was at Awash, thus defining the economic frontier which designated Harerge as 'a free trade zone within a socialist state.'

The government adopted this policy for two main reasons. First, in the late 1970s, Harerge was emerging from war devastation, and contraband trade was seen as a source of employment for young men and women who otherwise would have engaged in prostitution, thievery and other socially unacceptable activities. Secondly and more importantly, there was a concern that, if the unofficial trade was prohibited, the Issa who sold khat and used the money to purchase rice, sugar, tea and other necessities would retaliate by damaging the railway line, thereby causing graver economic harm than they would have inflicted simply by smuggling goods.[39]

Contraband trade brought to Harerge economic changes that had significant repercussions on regional employment and income. By the end of the 1980s Harerge's rapidly expanding khat economy, which, to a significant extent, financed the import of contraband consumer goods, had incorporated a significant portion of the province's population. The segment of the population that gained part or all of its income from khat-related activities was larger than that employed by the few factories that existed in the region. Those who depended on the khat economy and the related parallel market activities were khat cultivators; truck-owners and truck drivers; trimmers who prepared khat for export; smugglers; and retailers.

[37] Ibid., pp. 10–11.

[38] HPA-KF.2, Zelleke Beyene, Chief Administrator of Harerge, Minutes of the High-Level Committee which studied the Problem of Contraband Trade, February 20, 1980.

[39] Tesfamichael, *State of Illicit Khat Trade*, pp. 19–20.

By the late 1980s contraband trade had become an integral part of Harerge's economy. Some argued that the parallel market activities were antithetical to economic development because they denied government the revenue it deserved. On the contrary, the phenomenon appears to have made surprising contributions to Harerge's economy by providing employment to people who would otherwise crowd cities and cause social problems and by serving as a social lubricant in easing tensions stemming from state interventions. There was more to be gained from the positive contributions of contraband trade to the socioeconomic well-being of the region than the revenue it might have denied the government.

Ten

Conclusion
The dilemma of khat-based development

To a varying degree, khat has always occupied an important place in Harerge's social and cultural institutions. It is important to recognize that the production and distribution of khat have led to conflict between traders, ethnic groups, farmers and officials. Hareri farmers jealously kept the shrub from spreading to the Oromo until the late nineteenth century, when they hired the latter to work their khat orchards. In the early twentieth century government officials attempted to eradicate khat in order to 'liberate farmers from its evil influence.'[1] Despite their determined effort, khat has been Harerge's most profitable cash crop and its consumption the most pervasive social habit since the 1940s. Governments, especially those on the importing side, have favored controlling the industry in the hope of attaining dominance over their respective economies and the population's behavior. The efforts collapsed in the face of a stiff resistance put up by farmers and traders.

The rapidity with which khat consumption expanded in the Red Sea region was made possible by the construction of transport networks (roads, railway line, an international airport at Dire Dawa), which were originally intended to develop export crops other than khat and enhance

[1] The use of khat has been a source of controversy in almost all of the countries where the leaf is used. There have been efforts to prohibit its use. Tracing the khat debates during the last half-century is a subject that will provide an occasion to survey the role of the leaf in Ethiopia's sociopolitical history, but that is a separate project to be studied at another time. Those interested in pursuing the matter are referred to the khat debates in Aden, Somaliland Protectorate and several international fora, and the many documents in the following files at the PRO in London: CO 323/1885/7, Opium and Dangerous Drugs, Inter-Departmental Opium Committee: use of khat, 1945–1946; CO 323/1924/4, Somaliland: use of khat and other drugs, 1950; CO 323/1924/5, Somaliland: use of khat and other drugs, 1951–1952; CO 822/1824, Use of the drug Khat (or Qat or Kat) in East Africa, 1957–1958; CO 1015/2050, Use of drug Qat or Khat in Aden and British Somaliland, 1957–1958; CO 1015/2051, Use of drug Qat or Khat in Aden and British Somaliland, 1958–1959.

agricultural productivity. Modern transportation helped to open up foreign markets for the leaf, allowing farmers quickly and decisively to respond to declining production, falling coffee prices, demographic pressure and land scarcity and state intervention. After analyzing available goals and potential risks, they pursued strategies appropriate to rapidly shifting political and economic contexts. Despite numerous obstacles, they expanded and intensified khat cultivation and provided a steady supply of the leaf to meet the needs of consumers in the Horn of Africa and southern Arabia. Constantly striving to outmaneuver unfair and intrusive policies, khat farmers have in different ways powerfully shaped the agrarian structure of Harerge.

The seemingly insatiable demand for khat in the countries of the Horn of Africa and the Arabian Peninsula has been responsible for driving up both the price and levels of khat production. In 1994, on a normal day, an estimated 225 buyers came to the evening market and 350 persons to the early morning market at the AwWoday market center. Each buyer collected various quantities of khat ranging between 60 and 700 kg. Calculated on the basis of these figures, an average 218,500 birr ($43,720 in 1994) worth of khat changed hands at the marketplace. This figure translates to 79.8 million birr or $16 million spent annually on purchasing khat in AwWoday only. Production, likewise, has reached staggering proportions, increasing from an annual average of 70 metric tons in 1961 to 9,644 in 1985.[2]

Not all of this khat was legally exported or consumed within the country. Correlated with the khat industry was a flourishing parallel market, often stigmatized by authorities as illegal and illegitimate. The khat economy fed other illegal activities, such as smuggling and money laundering, to the point that the unofficial sector in Harerge by the early 1990s was more important than the official. Harerge's parallel markets were estimated to involve approximately 1 billion birr, and khat smuggling accounted for nearly a third of it.[3]

The conversion of khat to a commodity of high, concentrated value created an opportunity for Harerge's farmers to improve their standard of living when most rural Ethiopians were experiencing a decline. For example, in Harerge fewer people were seen in threadbare and patched clothes; many wore imported jeans and shoes. Electrification had reached some homes in rural areas, and appliances such as tape players and television sets were found in small-town shops and even in some farmer homes. For the most part these commodities entered Ethiopia as contraband goods and their sale served to repatriate some khat profits.

Khat-generated money was displayed in ostentatious ways. There was plenty of money for celebrating social occasions in manners hitherto unseen in Harerge's rural areas. By the late 1980s many farmers had

[2] Ministry of Foreign Trade, Branch Office for Eastern Ethiopia, 'Foreign Trade Activities in Harerge Province' (Dire Dawa, 1992), pp. 6–8.

[3] OPCWH, *Contraband Trade in Western Harerge*, pp. 6–9; Abebe *et al.*, *Illicit Trade in Eastern Ethiopia*, pp. 67–75.

become owners of minibuses, pick-up trucks and taxis. The staple food of ordinary people also changed. Instead of the traditional *lafiso* (bread) made of sorghum, people ate rice, pasta and canned food. Some families reabsorbed people who could not succeed in cities, hiring them as drivers for their commercial vehicles. Farmers generally radiated self-confidence, although some were unsure of the sustainablity of the market for the leaf. The majority of farmers benefitted from the high price of their khat, the only commodity whose price kept ahead of inflation and which, during the command economy years of the late 1970s and 1980s, escaped government price controls. The change in the farmers' standard of living is captured in Adem Aliyu Jaylan's life. He remembers:

> As khat prices increased, my daily wages rose proportionally. For the last four years [1990–4], I have been paid 200 birr for a day's worth of work by my employer, that is, the exporter who extended the money for purchasing khat. Sometimes I sell my own khat to the exporter, the khat that I purchased with my own money, and make more money.
>
> Today I own three vehicles – one is my private family car, two are used for commercial purposes. All three are driven by my three sons who possess legal driver licenses. Apart from my own residence, I own 30 houses that I rent out to tenants here in AwWoday. I also own two shops (stores) carrying consumer commodities. This business of khat has completely changed our lives, especially since 1975 E.C. (1982–3).
>
> I remember the days when there were a few radio sets in town. I used to go to the houses of those who own them to listen to radio programs. Today I have a televison set and an automatic telephone in my house. My father lived in a hut. I live in a very modern house with my 18 children whom I am raising without any financial problem. You asked me to compare my life to that of my father. I really can't say my father had a life, let alone a decent life. I can't compare our life with that of my father. It's night and day. [4]

The investment of wealth in vehicles was an ideal choice in an economically and politically risky environment. In an economic context marked by high inflation and lack of investment opportunities, the strategy of buying vehicles functioned as a shield against accelerated erosion of the purchasing power of local currency. Compared to other options, investing in vehicles was especially attractive for farmers because it was a reliable means of generating and accumulating additional wealth. Farmer-owned commercial vehicles were visible on Harerge's main highways, where they transported passengers and agricultural products including khat, to and from Dire Dawa, Harer, Jijjiga and other towns.

For much of the last fifty years the khat industry represented a lifeline for many of the rural and urban poor who have employment in some aspect of the business. While farmers became increasingly dependent on khat production for their survival, a significant number of the urban unemployed benefitted from engaging in the distribution of the leaf. The

[4] Interview: Adem Aliyu Jaylan, AwWoday, January 19, 2000.

trade played an important social role for women, who participated in the retailing of the substance in liberating them from dependence on their husbands. Most of the small retailers who purchased khat at AwWoday market and sold it in small bunches in nearby towns were women. Farmers' wives sold the poor-quality '*faaqaa*' variety directly to consumers and used the proceeds to purchase merchandise for themselves and their families. In interviews, many of these women often affirmed that khat made them financially independent.[5]

Petty employment in buying, transporting and selling khat leaves in legal as well as illicit markets has been a traditional occupation for many Harerge residents living at subsistence level. In 1979, when the government permitted some persons to purchase and transport up to 100 birr worth of khat leaf to Djibouti, there was a weekly traffic of at least 4,000 people into the khat-purchasing areas. The low price offered in official markets caused many farmers and traders to opt for the alternative illegal markets and, by the 1980s, the amount of khat smuggled into the neighboring territories had reached almost half the officially exported amount and was far greater in terms of value.[6]

Growth in khat's economic importance presented new economic opportunities for social groups not connected with the production of khat, particularly commercial intermediaries already engaged in a range of market activities. This class of traders obtained khat from producers, smuggled the produce into neighboring countries, and used the proceeds to supply contraband consumer durables and even basic foodstuffs. A substantial portion of the high profits derived from smuggling khat came to play an influential role in shaping the character of the regional economy as much of the capital was used to finance a parallel market that created employment opportunities in the region for more people than the public and manufacturing sectors combined.

The commodification of khat perhaps benefitted farmers directly in providing a temporary solution for the apparently intractable problem facing Harerge's agriculture – shortage of arable land. In Harerge, particularly in the eastern and central districts, as much as 90 percent of the land is cultivated, and the average household land size had dwindled to 0.84 hectares in 1990 compared to 1.1 hectares in 1980 and 1.5 hectares in 1965.[7] As we have seen in Chapter 6, in the 1980s the average yield per hectare for sorghum was 19 quintals. Since it is impossible to survive on this quantity of sorghum, farming households found it imperative to maximize the productivity of the available land.

5 Interviews: Kimya Ahmed, Ganda Limay, Haramaya, June 2, 1994; Nuriya Mohammed, Ganda Kollobo, Kombolcha, June 7, 1994; Kemero Ali, Haramaya, May 27, 1994.

6 World Bank, *Ethiopia: Export Action Program*, p. 47.

7 P. Poschen, 'An Evaluation of the *Acacia albida*-based Agroforestry Practices in the Hareghe Highlands of Eastern Ethiopia,' in P.K.R. Nair (ed.), *Agroforestry Systems in the Tropics* (Dordrecht, 1989), pp. 385–400; Badeg Bishaw, 'Determining Options for Agroforestry Systems for Hareghe Highlands, Ethiopia,' Ph.D. dissertation, Department of Forestry, Oregon State University, 1993.

By the early 1990s khat had become part of the diversified cropping system that fitted well into both the environment and the agricultural cycle of small-farm households without being too demanding of labor. Harvested at least three times a year, it provided a regular source of income to meet household expenditure. Khat also consistently maintained price advantages over other crops. In the 1980s, it yielded three times the returns on sorghum, ten times the returns on coffee and nearly twice the returns on low-input management potato. By the mid-1990s this price differential had widened further, with per hectare profit from khat exceeding all other crops, making government-sponsored substitution programs unattractive and unsuccessful.[8]

The most enduring impact of khat was the fact that its trade provided badly needed capital for individuals to leave their farms and look for nonagricultural opportunities elsewhere. The transport businesses, brick factories, filling stations and small shops that sprouted all over Harerge highlands were results of khat-generated capital. These ventures enabled farmers to extricate themselves from land that over the years had become extremely scarce and diminished the possibility of many becoming smallholder, self-sufficient farmers. In this regard, khat did what the policy planners would have never imagined they could do for farmers, that is, moving them to nonagricultural occupations, thereby easing demographic pressure on land and agricultural resources. It is interesting to note that farmers pursued diversified occupational strategies to cope with the challenges of agriculture whereas policy-makers sought to fix the existing problems of agriculture and perpetuate smallholder farms.

The commercialization of Harerge's agriculture did not lead to the spread of capitalist relations of production and the transformation of farmers into wage laborers. Household production remained the dominant form of economic activity in the region. Households essentially secured social reproduction from the cultivation of the leaf, but, despite the appearance of conspicuous consumption, the production process itself was not capitalized. The marketing process worked to the disadvantage of the producers. Khat prices rose exponentially, but much of the capital thus acquired contributed to circuits of primitive accumulation elsewhere. The production of khat in Harerge was characterized by the draining of profit away from the producer. In the early 1990s the disproportionate distributional rewards resulting from involvement in the khat industry were increasingly apparent in the Harerge region. For the majority of the primary producers and traders in rural and urban centers, access to basic services was heavily restricted and consumption needs were met only by involvement in illicit activities. Despite rising prices, most khat producers did not enjoy the benefit of clean drinking water or all-encompassing rural electrification schemes, and health posts, schools and roads in the region were seriously deficient in number and quality.

[8] At the time of my research in 1994, I met farmers who had sold 1 kg of khat for 1,100 birr that year.

Despite the tangible improvement that the khat boom brought to the lives of hundreds and thousands of the rural and urban poor, it is not possible to state with confidence that it presented a suitable route for regional development. The market for the product was limited and precarious. One cannot expect that the consumption of khat, which has been confined to the Red Sea region for many centuries, will expand and develop to an international market. Even the existing market was fraught with serious problems emanating from policy changes in the receiving countries, and is likely to cause a major economic problem for the producers in Harerge.

The prosperity that khat cultivation generated has been a two-edged sword. While it has provided some concrete, short-term benefits to farmers and other segments of Harerge's impoverished unemployed and a certain degree of respite from economic devastation and political turmoil, it did not open an avenue leading to structurally sound and sustainable development. Harerge's farmers have managed to create an agrarian system that can sustain an extremely high population, the density of which, in some places, reaches 400 people per square kilometer. However, if one pays close attention to the sequential and cumulative impacts, the relations between past and present as well as local and broader options, actions and strategies, the prognosis of Harerge's agriculture is not one of a stable future. Harerge's high population density has resulted in an expansion of cultivated lands, at the expense of grazing lands, thereby decreasing the quantity of fodder for animals. In addition, the development of cash crops and the consequent commercialization of Harerge's agriculture have reduced the land cultivated with maize and sorghum, and, therefore, the quantity of fodder that can be obtained from these two crops.

The shortage of fodder, in turn, has led to reductions in the number of livestock a household can keep and, hence, in the amount of manure that may be collected. Customarily, the limited amount of manure is applied in priority to cash crops. Even though food crops have always occupied much of the cultivated land, chemical fertilizer was used only sparingly owing to high cost. The disproportional allocation of manure has caused a decrease in the humus content of the soil and the productivity of the land. In many places, erosion has washed away the topsoil and farmlands have become impoverished. Having nearly exhausted its topsoil, Harerge's agriculture has come to experience a condition of extreme fragility. If no solution is found, it is likely that the region's agriculture will face in the near future a very serious ecological crisis.

The high population density has also resulted in the diminution of household land size and a necessity to intensify the use of the available arable land. Apart from extensive cultivation, reclamation of new lands and reduction of fallow periods and grazing lands, farmers achieved their goals by increasing the proportion of high-income cash crops relative to food crops. Khat and potato were prominent cash crops in the agrarian system, selected by farmers because of the fact that both crops could be sold or consumed. In this regard, potato has a crucial role to play in the overall balance of the agrarian system. For example, when the supply of grain from outside Harerge was sharply reduced in the summer of 1984, farmers and urban-dwellers overcame the shortfall by supplementing their diet with potato. The farmer determined the proportion of land each crop occupies, but competition from khat for land with food crops grew in intensity as its return on investment increased with rising prices. The dilemma was whether to increase the proportion of cash crops and risk becoming totally dependent on the market or to continue to grow food crops and provide food security for the farming household. Each farmer made the decision regarding the proportion of khat and food crops to be planted on the basis of family needs, farm-land size and market conditions.

Although farmers chose the crops they planted, external factors also influenced their decisions. For food crops to play the major role of security, farmers had to cope with climatic variability and with such changing conditions of production as pests and disease attacks and market fluctuation, as well as unexpected change from one year to another. In the face of these uncertainties, farmers tended to minimize the consequences of variability for the overall environment of production and to maximize the production of each crop. The avoidance of risks and the attainment each year of a sufficient and stable production proved to be as important for farmers as the profit from selling cash crops.

To achieve their objectives, farmers used crops that respond to environmental changes differently, relied on crops that provided them cash at different periods of the year and allowed them to minimize the effects of failure on their food supply. Their objective was to adapt themselves to any unforeseeable change, be it a drop in prices or a crop failure. Markets alone did not determine the cropping pattern.

Analysis of regional agriculture shows that the variety of crops in Harerge was not just a matter of happenstance or the adaptability of those crops to the agroecological environment. The types of crops found in the region were chosen and combined by the farmers themselves because each crop played a particular role in the overall agriculture. The combination of cultivated crops and the proportion of each crop grown has been the direct result of the farmers' choice and the logic of production they followed.

If one looks closely at the case of Harerge, the distinction between cash crops and food crops is rather blurred, because farmers have chosen to produce crops that can be consumed or sold. Accordingly, farmers have chosen to produce khat and vegetables and to buy grain. In the 1980s the *derg*'s policy of holding grain prices low to benefit urban consumers encouraged farmers to grow more profitable cash crops and purchase less expensive cereals. In the process, farmers came to view khat as their 'miracle crop,' which enabled them to survive hard times. Regardless, they remained keenly aware that the choice they had made was not a viable, long-term solution for the region's agriculture.

This study has shown how independent-minded farmers cope with environmental, political and economic adversities. Government policies that centered on prescribing the strategies farmers must follow, without taking into account the views of farmers themselves, have not succeeded in bringing about agrarian transformation. It seems that Harerge's farmers have identified a successful rural development strategy. Most use their cash wealth to enter nonfarm life, but they have limited knowledge of the opportunities that are available. Policy-makers could benefit the farmers by helping them identify and invest their own money in workable, nonfarm economic activities rather than insisting on the perpetuation of smallholder agriculture. Harerge's agricultural and developmental problems cannot be solved by allocating 'land to the tiller,' for it is no longer available. The popular political motto of yesteryear, which galvanized the masses of Ethiopia into launching a radical revolution in 1974, has outlived its political usefulness, much less its practical viability as a policy choice. The only solution to agricultural problems resulting from land scarcity is what Harerge's farmers have already figured out: shift to nonfarm economies.

Appendix A

Coffee & khat exports from Harerge, 1936–91

Year	Coffee		Khat	
	Volume (tons)	Value ('000 birr)	Volume (tons)	Value ('000 birr)
1936	11,883	12,153	—	—
1937	7,052	12,031	15	7
1938	15,228	18,326	26	13
1946	—	—	26	14
1947	—	853	81	43
1948	—	818	204	147
1949	—	1,082	186	183
1950	—	1,088	368	246
1951	—	—	470	723
1952	—	—	768	1,44
1953	—	—	1,176	1,806
1954	—	—	1,228	5,058
1955	—	—	1,522	6,556
1956	—	—	1,618	6,996
1957	—	1,229	821	3,601
1958	—	971	1,386	4,782
1959	—	979	2,300	7,317
1960	—	—	2,149	6,900
1961	—	—	2,690	8,686
1962	—	—	3,856	10,444
1963	—	—	4,798	12,541
1964	—	—	1,966	5,051
1965	—	—	—	—
1966	—	—	799	2,170
1967	—	—	854	2,921
1968	—	—	842	3,003
1969	—	—	1,212	5,009
1970	7,238	18,897	607	1,747
1971	5,358	12,283	1,931	4,042
1972	7,964	19,384	2,751	4,267
1973	7,934	20,453	2,059	3,838
1974	6,907	21,041	2,335	5,538
1975	5,816	17,667	2,504	5,425
1976	8,420	45,046	3,384	7,808
1977	4,109	49,207	747	5,335
1978	—	—	1,062	6,625
1979	2,355	16,213	1,997	13,978
1980	3,103	25,625	1,874	19,379
1981	1,254	11,057	2,581	28,480
1982	1,072	10,654	2,824	31,067
1983	2,260	18,646	2,805	30,854
1984	2,663	17,881	2,904	32,028
1985	604	6,152	2,804	31,424
1986	659	9,695	2,699	28,348
1987	804	11,355	2,863	32,157
1988	544	6,643	2,182	20,881
1989	807	8,461	2,868	36,334
1990	1,228	15,210	2,425	31,760
1991	1,281	12,577	1,184	15,228

Sources: PMAC, Customs & Excise Authority. Annual External Trade Statistics, 1960–1991; National Bank of Ethiopia, Dire Dawa Branch; FO 371/53465; *Bulletin of the Ministry of Commerce and Industry*, 13 (1964).

Appendix B

Glossary

Ato	A title, equivalent of 'Mr'
Awraja	Administrative unit; sub-province
Balamberas (Balam)	Military title; literally 'Commander of a fortress'
Bejirond	Honorary title given to treasury officials
Bitwoded (Bit)	A title of nobility; literally 'beloved'
Dejazmatch (*Dej.*)	Military title; literally, 'Commander of the gate.' Conferred on members of the Ethiopian aristocracy
Derg	Amharic term for committee. Used to refer to the military regime which ruled Ethiopia from 1974 to 1991
Fitaurari (*Fit.*)	Military title; literally, 'Commander of the vanguard.' Conferred on the nobility and commoners
Geber	Tribute, invariably paid in kind; tax
Grazmatch (Graz)	Military title, literally 'Commander of the left flank.' Conferred on commoners
Inderassie	Title of nobility; literally, 'representative "of the emperor"'
Kella	Toll-gate
Kuter gebir	Unsurveyed or unmeasured land, usually held by farmer-operators
Lij	Royal title; literally, 'child'
Qalad	A piece of rope used for land measurement. Also denotes a measured plot of land
Qegnazmatch (*Qegnaz*)	Military title; literally, 'Commander of the right flank.' Conferred on commoners
Ras	Military title; literally 'head.' The highest military rank given to the nobility and commoners
Shibetta	A designation of a plot of land found in Harerge
Woreda	Administrative unit; district

Appendix C

Selected informants

Abdi Bakar (Hajji), 60, April 6, 1994. Khat farmer, Ganda Lucha, Haramaya

Abdo Shanqo, 52, May 19, 1994. Khat farmer/employee Haramaya University of Agriculture, Bate Worke PA, Haramaya

Abdulahi Ahmed, 64, May 17, 1994. Khat farmer, Ganda Issa, Bate Worke PA, Haramaya

Abdulahi Mumed, 70, May 18, 1994. Retired khat farmer, Ifa PA, Haramaya

Abdulahi Mussa, 62, June 9, 1994. Khat farmer, Ganda Faldhaana, Kadir PA, Kombolcha

Abdulahi Shebo, 70, May 25, 1994. Coffee/khat farmer, Ganda Lucha, Deda PA, Haramaya

Abdule Yusuf, 64, June 8, 1994. Khat farmer, Ganda Lencho, Kayrensa PA, Kombolcha

Adem Aliyu Jaylan, 61, AwWoday, January 19, 2000

Ahmed Abdulahi, 50, June 7, 1994. Khat/coffee farmer, Ganda Saqa Sharifa, Samte PA, Kombolcha

Ahmed Amit, 61, May 25, 1994. Khat farmer, Ganda Gole, Haji Walta'a PA, Haramaya

Ahmed Tewfiq Mohammed (*Balambaras*), 68, March 23, 1994. Member of Parliament (1958–62), *woreda* administrator of Garamulata, later of Chercher, Dire Dawa and later *awraja* administrator of Garamulata, Harer

Alemayehu Woyessa, 60, AwWoday, January 19, 2000

Avidas Terzian, c. 80, August 8, 1994. Addis Ababa. An Armenian businessman and community leader. Son of Sarkis Terzian, one of the first Armenians to locate in Ethiopia in the late nineteenth century

Beshir Sheik Abdi (*Fitaurari*), 68, March 24, 1994, Harer. *Woreda* administrator of Ferfer, *awraja* governor of Dagahbour and later of Jijjiga, Deputy Governor of Harerge region (1975–91), Harer

Dawd Abdo (Hajji), 60, Dire Dawa, March 26, 1994. Businessman, former khat exporter, member of the khat exporting company in the 1950s

Duressa Feyissa, 73, February 24, 1994, Dire Dawa. Worked for the French with the RR company since 1941. Moslem and Oromo (from Wollega). Also participated in the illegal trade during the 1970s. Well versed in events since 1941

Girma Andarge, 84, Dire Dawa, February 23 and 27, 1994. Metal worker, worked with the Italians during the occupation in Harer. Remembers most things that happened as far back as the time of the regency

Ibrahim Dawe, 40, May 30, 1994. Khat trader (*kebkebi*), AwWoday khat marketing center

Ibrahim Yusuf, 60, June 9, 1994. Khat farmer, Ganda Feline, Kafir PA, Kombolcha

Ismael Mohammed, 60, June 7, 1994. Khat farmer, Ganda Kollobo, Igagena PA, Kombolcha

Luye Said, 64, May 26, 1994. Khat farmer, Ganda Limay, Bate Worke PA, Haramaya

Mohammed Abdulahi Ogsaday, 74, February 22, 1994. Coffee exporter. Began exporting coffee from Dire Dawa in 1944

Mohammed Abdurahman (Hajji), 60, March 26, 1994. Fresh out of Alliance Française school in Dire Dawa, he joined the khat exporting business in 1951. Left khat exporting when CHATHA was established, protesting government meddling in the trade. Employed by the Djibouti–Ethiopian railway as Secretary General in 1961 until his retirement in 1989. He knows well the details of khat trade for the time he was involved in it

Mohammed Yusuf, 58, May 25, 1994. Coffee/khat farmer, Ganda Lucha, Deda PA, Haramaya

Mume Kiko (Hajji), 59, May 20, 1994. Khat farmer, Haji Walta'a PA, Bate Haramaya

Mumed Ahmed Umer, 72, May 17, 1994. Khat farmer, Ganda Issa, Bate Worke PA, Haramaya

Mumed Munan, 71, May 24, 1994. Coffee/khat farmer, Ganda Lucha, Deda PA, Haramaya

Mumet Abdule, 75, May 25, 1994. Coffee/khat farmer, Ganda Lucha, Deda PA, Haramaya

Mussa Botto, 88, May 23, 1994. Khat farmer, Ganda Gole, Mula'ata PA, Haramaya

Nuriya Mohammed, 46, June 7, 1994. Khat farmer's wife/transport vehicle operator and kiosk owner, Ganda Kollobo, Igagena PA, Kombolcha

Rayima Sali, 75, May 23, 1994. Khat farmer's wife, Ganda Gole, Mula'ata Pa, Haramaya

Taddesse Debele (Captain), February 23, 1994, Dire Dawa, General Manager of the Khat Exporters Cooperative of Ethiopia (KECE)

Yusuf Hammo, 75, May 18, 1994. Retired khat farmer, Ifa PA, Haramaya

On May 26, 1994, a group of 20 khat farmers and their wives were interviewed at Ganda Lucha, Bate Worke PA, in Haramaya

Bibliography

Archival material

Ethiopia

EASTERN HARERGE ADMINISTRATIVE ZONE (EHAZ)
Documents Relating to Khat, Folder 1 (EHAZ-KF.1) 1979–1986

HARERGE PROVINCIAL ARCHIVES (HPA)
'About the administration of land,' *Minutes of the General Assembly Debate on the Question of Land Tenure*, 1950
Correspondence Relating to Khat, Folder 1 (HPA-KF.1) 1960–1989
Correspondence Relating to Khat, Folder 2 (HPA-KF.2) 1970–1986
General Correspondence, Folder 1 (HPA-GC.1) 1958–1981
General Correspondence, Folder 2 (HPA-GC.2) 1964–1974

MINISTRY OF FOREIGN TRADE (MFT) (DIRE DAWA)
General Correspondence, Folder 1 (MFT-CF.1) 1979–1991
Annual Reports, 1981–1991
Correspondence Relating to Contraband, Folder 2 (MFT-CF.2) 1982–1991
Report on the State of the Province's Foreign Trade 1989–1990
Report on Foreign Trade Activities of the Province (Harerge), 1991

MINISTRY OF INTERIOR (MOI)
Annual Report from Harerge Province (RHP) 1935–1936, 1970, 1977, 1984

CUSTOMS AND EXCISE AUTHORITY
Annual External Trade Statistics, 1960–1991

Great Britain

The Public Record Office, London
Foreign Office 368/1688
371/146597, 371/126868, 371/16996, 371/16101
371/53561, 371/53465, 371/53461
371/80232
401/10, 401/24
401/30, 401/34
403/255, 403/298
Colonial Office 323/1885/7

323/1906/5, 323/1924/5, 323/1924/6, 323/1906/6, 323/1906/7
324/1924/4, 324/1924/5
822/1824
1015/2050, 1015/2051, 1015/2052

United States

Department of State Decimal Files
SD 875.00/7-957; SD 875.00/4-557; SD 875.00/7-2456

France

Archives Nationales de France Dépôt d'Outre Mer, Aix-en-Provence
Bulletin de Comité d'Afrique (BCAF), 1899–1907. No author or title
Hachette, René, 'L'Importance de la Côte Française des Somalis,' *Revue Géographique et Industrielle de France*, 2 (1936), in AOM, Carton 341 Dossier 1
Importations, Colonie de la Côte des Somalis, AOM, Carton 342 Dossier 7 and Carton 346 Dossier 11

Official publications

Aden Colony, *Proceedings of the Legislative Council's 11th Session*, Printed by the Government Printer (Aden, 1957)
Aden Colony, *Report of the Qat Commission of Inquiry*, Printed by the Government Printer (Aden, 1958)
Bulletin of the Ministry of Commerce and Industry 13 (1964)
Central Statistical Office, *Statistical Abstract,* 1972
Central Statistical Office, *Results of the National Sample Survey, Second Round: Land Area and Utilization.* Vol. 5 (Addis Ababa, 1975)
The Consolidated Laws of Ethiopia, vol. 1, as quoted in *Meskerem,* monthly magazine, 16 (1984)
House of Commons Sessional Papers, *Reports for the Year 1898–1900 on the Trade of Adis Abbaba and Harar*, 92 (1900)
House of Commons Sessional Papers, *Report for the Year 1905–06 on the Trade of Abyssinia* (Cd. 3283–8), 88 (1907)
Journal Officiel, 5 Avril, 1957
Mackereth, G., *Economic Conditions in Ethiopia*, Department of Overseas Trade. No. 507 (1932)
Ministry of Agriculture (MOA), *General Agricultural Survey Preliminary Report, 1983–84, Harerge*, vol. 2. Addis Ababa, 1984
Negarit Gazetta 10, 1935 E.C. (1942)
Somaliland Protectorate Annual Report, 1930. London, 1931
Somaliland Protectorate Annual Report, 1936. London, 1937
US Department of Commerce, *Ethiopia: Commercial and Economic Survey*, Special Consular Reports, No. 81 (1918)
US Department of Commerce, *Ethiopia: Commercial and Economic Survey*, Trade Information Bulletin, No. 476. Washington, D.C., 1927

Unpublished essays, theses, dissertations & manuscripts

Abebe Kebede, Abebe Mekonnen, Mesfin Shiferaw, Samia Yusuf and Teferi Tilahun, *The Causes and Characteristics of Illicit Trade and Future Trends in Eastern Ethiopia* (Amharic). Dire Dawa, 1989

Abraham Gebremariam, Abebe Kebede and Yoseph Gebrehiwot, *Khat Exporters' Cooperative of Ethiopia: Progress and Prospects* (Amharic). Dire Dawa, 1990

Addis Tiruneh, 'Gender Issues in Agroforestry.' Paper Prepared for Workshop on 'Land Tenure and Land Policy in Ethiopia,' Institute of Development Studies. Addis Ababa, 1994

Aehling-Kriesel, D. and Kifle Lemma, *Farming Systems Development Survey Report.* Ministry of Agriculture, People's Democratic Republic of Ethiopia, Field Document TCP/ETH/6659, 1989

Badeg Bishaw, 'Determining Options for Agroforestry Systems for Harerghe Highlands, Ethiopia.' Ph.D. diss., Department of Forestry. Oregon State University, 1993

Brooke, Clarke, 'Settlements of the Eastern Galla, Hararge Province, Ethiopia.' Ph.D. diss., Department of Geography. University of Nebraska, 1956

Darby, W.J. *et al.* 'Khat (or Chat).' In *Ethiopian Nutrition Survey. A Survey by the Interdepartmental Committee on Nutrition for National Defense.* Washington, D.C., 1959

Dejene Aredo and A. Abduraheman, 'A Study of the Contraband Trade in Eastern Ethiopia.' Unpublished Mss. Department of Economics. Addis Ababa University, n.d.

Demissie Gebre Michael, *The Kottu of Harerge: An Introduction to the Eastern Oromo.* Imperial Ethiopian College of Agriculture Experiment Station Bulletin No. 68. Dire Dawa, 1974

Desta Degefu and Masresha Yimer, *Conditions of Social Security in Harerge* (Amharic). Prepared for Eastern Ethiopia Zone Planning Office. Harer, 1986

Eastern Ethiopia Ministry of Foreign Trade, *Report on the State of the Province's Foreign Trade* (Amharic). Dire Dawa, 1989

Eastern Ethiopia Zone Planning Office, *Bulletin of the Socio-Economic Conditions in the Zone* (Amharic). Harer, 1988

Eastern Ethiopia Zone Planning Office, *A Comprehensive Report on Contraband Trade Activities and the 'Taiwan' Contraband Market* (Amharic). Harer, 1988

El Mahi, T., 'A Preliminary Study on Khat Together with the Institutional History of Coffee as a Beverage in Relation to Khat.' World Health Organization, Regional Office for the Eastern Mediterranean. EM/MENT/29/4. March 1962

Ferman, L.A., L. Berndt and E. Selo, 'Analysis of the Iregular Economy: Cash Flow in the Informal Sector.' Report to the Bureau of Employment and Training, Michigan Department of Labor, Institute of Labor and Industrial Relations, University of Michigan/Wayne State University. Ann Arbor, Michigan, 1978

Getachew Eshete, 'Contraband Trade in Ethiopia: Major Highlights.' Paper Presented at the Seminar on Contraband Trade. Addis Ababa, 1995

Girum Tamirayehu, 'Coffee Production Problems and Cultural Practices in Harerge.' In Mesifin Ameha, ed., *Proceedings of the First Workshop on Production Constraint Assessment of Coffee, Tea and Spices (CTS).* Jimma, 1991

Hagos Gebre, 'Contraband Trade in Ethiopia' (Amharic). Prepared for the Office of the National Committee for Central Planning. Addis Ababa, 1988

Hasen Ali, 'A Short Biography of Bejirond Takla Hawariat.' B.A. essay. Department of History, Addis Ababa University, 1985

Kidanemariam Gebreegziabher and Abay Asfaw, *A Study on the State of Illicit Trade in Harerge Region* (Amharic). Prepared for Office of Central Planning for Eastern Ethiopia. Harer, 1988

Lakew Birke, *A Study of the Development Impact of the Miesso–Dengego Road*. Department of Agricultural Economics and Business, College of Agriculture, Haile Sellassie I University. Alemaya, 1973

Lakew Birke, *A Survey of the Marketing Methods and Facilities in Alemaya District*. College of Agriculture, Haile Sellassie I University. Dire Dawa, 1974

Langlais, C., M. Weill and H. Wibeaux, *Farming Systems Research: Preliminary Report and Future Programme*. Alemaya University of Agriculture, Alemaya, 1984

Lemma Arity, 'A Regional Study of Hara Maya Based on the Studies of the Alemaya College of Agriculture.' B.A. essay. Department of Geography, Haile Sellassie I University, 1969

Milkessa Wakjira, 'An Economic Analysis of the Development for the Alemaya Lake Sub-watershed with Special Reference to Soil and Water Conservation Methods.' M.Sc. thesis. Department of Agricultural Economics, Alemaya University of Agriculture, 1989

Miller, Leonard F., *Input–Output Data for Chat and Sorghum Production, Harer Highlands, Ethiopia*. Imperial Ethiopian College of Agriculture and Mechanical Arts, Alemaya, 1965

Ministry of Foreign Trade, 'Guidelines for Regulation of Import–Export Trade.' Dire Dawa, 1988

Ministry of Foreign Trade, Branch Office for Eastern Ethiopia, 'Report on Ways to Control Contraband Trade at the Train Station and Safeguard the Train' (Amharic). Dire Dawa, 1989

Ministry of Foreign Trade, Branch Office for Eastern Ethiopia, 'Report on the State of the Province's Foreign Trade.' Dire Dawa, 1989

Ministry of Foreign Trade, Branch Office for Eastern Ethiopia, 'Foreign Trade Activities in Harerge Province.' Dire Dawa, 1992

Ministry of Interior, 'The Demand for and Distributional Difficulties of Consumer Commodities in Border *Awrajas* and Recommended Solutions,' A report to the Ministry of Interior (Amharic), Addis Ababa, 1984.

Miran, Jonathan, 'L'occupation Egyptienne de Harar, 1875–1885.' Mémoire de Diplôme de Recherche et d'Etudes Appliquées Est-Africaines et Malgaches. Institut National des Langues et Civilisations Orientales, 1983

Mohammed Hassen, 'Menelik's Conquest of Harar, 1887 and its Effect on the Political Organization of the Surrounding Oromo up to 1900.' *Working Papers on Society and History in Imperial Ethiopia: The Southern Periphery from the 1880s to 1974*. Edited by D. Donham and Wendy James, African Studies Centre, Cambridge, 1987

Muhammad Hassan, 'The Relation between Harar and the Surrounding Oromo between 1800–1887.' B.A. essay. Department of History, Haile Sellassie I University, 1973

Negatu Wolde Giorgis, *Planning and Implementing the Development of Harerge Province* (Amharic). Harer, 1960

Office of the Planning Committee for the Western Harerge Administrative Area, *Guidelines to Control Contraband Trade in Livestock in Western Harerge Administrative Area* (Amharic). Asba Tafari, 1989

Risoud, Bernadette, *Creation and Development of a Producer Cooperative: Innovations and*

Interactions: The Case-Study of Tula in Harerge Highlands. Farming Systems Research, Alemaya University of Agriculture. Alemaya, 1988

Risoud, Georges, *Evolution of Peasant Agriculture in the Eastern Harerge Highlands: Development Prospects.* Farming Systems Research, Alemaya University of Agriculture. Alemaya, 1987

Shiferaw Bekele, 'The Railway, Trade and Politics: A Historical Survey, 1897–1935,' M.A. thesis. Department of History, Addis Ababa University, 1982

Takla Hawariat, 'The Autobiography of Fitaurari Takla Hawariat.' Department of History, Addis Ababa University, n.d.

Tesfamichael Gebremedhin, *The KEAE's Chief Inspector's Report on the State of Illicit Khat Trade along the Djibouti Railway Line* (Amharic). Dire Dawa, 1986

Tilahun Mekonnen, *Economic and Sociological Characteristics of Peasants Families at Alemaya, The Chercher Highland.* Department of Agricultural Economics and Business, College of Agriculture, Haile Sellassie I University. Dire Dawa, 1974

Tsehai Berhane Sellassie, 'Developments in Land Tenure in Harerge 1887–1935.' *Social Science Miscellanea 5.* Institute of Ethiopian Studies, Addis Ababa University, n.d.

ULG Consultants, *Coffee Marketing, Processing, Transport and Storage Study.* Vol. 3. Report Prepared for the Ministry of Coffee and Tea Development. London, 1986.

Weill, Marc, 'Marketing of Agricultural Products in Harar Highlands, Hararghe Administrative Region, Ethiopia, with Special Reference to Kombolcha *Woreda.*' Department of Agricultural Economics, Alemaya College of Agriculture, Alemaya, 1984

Wibeaux, Herve, *Agriculture in the Highlands of Hararghe, Kombolcha Area: Study of Six Farms.* Farming Systems Research, Alemaya University of Agriculture. Alemaya, 1986

World Health Organization, 'Medical Aspects of the Habitual Chewing of Khat Leaves.' WHO/APD/127 Rev. 1, 1964.

Articles

Abir, Mordechai, 'Brokerage and Brokers in Ethiopia in the First Half of the 19th Century.' *Journal of Ethiopian Studies* 3 (1965)

Allen, Catherine J., 'To be Quechua: The Symbolism of Coca Chewing in Highland Peru.' *American Ethnologist* 8 (1981)

Alles, G.A., M.D. Fairchild and M. Jensen, 'Chemical Pharmacology of *Catha edulis.*' *Journal of Medical Pharmacology and Chemistry* 3 (1961)

Amare Getahun and A. Krikorian, 'Chat: Coffee's Rival from Harar, Ethiopia, I. Botany, Cultivation and Use.' *Economic Botany* 27:4 (1973)

Andargachew Tesfaye, 'The Funeral Customs of the Kottu of Harer.' *University College of Addis Ababa Ethnological Society Bulletin* 7 (1957)

Archaya, S., and S. Madhur, 'Informal Credit Markets and Black Money: Do They Frustrate Monetary Policy?' *Economic and Political Weekly* 18:41 (1983)

Azam, Jean-Paul, and Timothy Besley, 'General Equilibrium with Parallel Markets for Goods and Foreign Exchange: Theory and Applications to Ghana.' *World Development* 17:12 (1989)

Bairu Tafla, 'Addictives in North East Africa: A Brief Survey of an Aspect of Nineteenth Century Cultural History. *Afrika und Ubersee* 64:2 (1982)

Bally, P.R.O., '*Catha edulis.*' *East African Medical Journal* 22 (1945)

Barker, J., 'Developments in Ethiopia's Road System.' *Geography* 59:2 (1974)

Befekadu Degefe and Tesfaye Tafesse, 'The Marketing and Pricing of Agricultural Products in Ethiopia.' In Siegfried Pausewang, Fantu Cheru, Stefan Brune and Eshetu Chole, eds, *Ethiopia: Options for Rural Development.* London and Atlantic Highlands, New Jersey, 1990

Beke, Charles, 'On the Countries South of Abyssinia.' *Journal of the Royal Geographical Society* 13 (1843)

Bhagwati, Jagdish, and Bent Hansen, 'A Theoretical Analysis of Smuggling.' *Quarterly Journal of Economics* 87 (1973)

Bhagwati, Jagdish, and T.N. Srinvasan, 'Smuggling and Trade Policy.' *Journal of Public Economics* 2 (1973)

Botta, P. E., 'Notice sur un voyage dans l'Arabie Heureuse.' *Paris. Musée Nationale d'Histoire Naturelle de Paris. Archives* 2 (1841)

Boucoiran, M., 'La Situation économique de l'Ethiopie.' *Renseignements Coloniaux et Documents* XI (1918)

Boulding, Kenneth, E. 'A Note on the Theory of the Black Market.' *Canadian Journal of Economics and Political Science* 13 (1937)

Britton, Everard, 'The Use of Qat.' *Geographical Journal* 93 (1939)

Bronfenbrenner, Martin, 'Price Control under Imperfect Competition.' *American Economic Review* 37 (1947)

Brooke, Clarke, 'The Rural Village in the Ethiopian Highlands.' *Geographical Review* 99:1 (1959)

Brooke, Clarke, 'Khat (*Catha edulis*): Its Production and Trade in the Middle East.' *Geographical Journal* 126 (March 1960)

Brooke, Clarke, 'The Durra Complex in the Central Highlands of Ethiopia.' *Economic Botany* 12:2 (1978)

Browning, Edgar, and William Patton Culbertson, Jr, 'A Theory of Black Markets under Price Controls: Competition and Monopoly.' *Western Economic Journal* 12 (1974)

Cancian, F., 'Risk and Uncertainty in Agricultural Decision Making.' In P.F. Bartlett, ed., *Agricultural Decision Making: Anthropological Contributions to Rural Development.* New York, 1980

Carothers, J. C., 'Mirra as a Cause of Insanity.' *The East African Medical Journal* 22 (1945)

Cassanelli, Lee, 'Qat: Changes in the Production and Consumption of a Quasi-legal Commodity in Northeast Africa.' In Arjun Appadurai, ed., *The Social Life of Things: Commodities in Cultural Perspective.* Cambridge, 1986

Caulk, Richard, 'The Occupation of Harar, 1887.' *Journal of Ethiopian Studies* 9:2 (1971)

Caulk, Richard, 'Harar Town and its Neighbors in the Nineteenth Century.' *Journal of African History* 3 (1977)

Chevalier, Auguste, 'Les Cat's d'Arabie, d'Abyssinie et d'Afrique orientale.' *Revue International de Botanique Appliquée et d'Agriculture Tropicale* 29 (1949)

Chibnik, M., 'Double-Edged Risks and Uncertainties: Choices about Rice Loans in the Peruvian Amazon.' In E. Cashdan, ed., *Risk and Uncertainty in Tribal and Peasant Economies*. Boulder, 1990

Coats, Peter D., 'Factors of Intermediacy in Nineteenth-century Africa: The Case of the Issa of the Horn.' In Thomas Labahn, ed., *Proceedings of the Second International Congress of Somali Studies*, Vol. II. Hamburg, 1984

Cook, Scott, 'The Obsolete Anti-Market Mentality: A Critique of the Substantivist Approach to Economic Anthropology.' *American Anthropologist* 68 (1966)

Cook, Scott, 'Our Obsolete Market Mentality.' In Karl Polanyi, *Primitive, Archaic and Modern Economies: Essays of Karl Polanyi.* Edited by George Dalton. New York, 1968

Corlett, Dudley S., 'Notes on the Products of Harar Plateau, Abyssinia.' *The Poona Agricultural College Magazine* (1917)

Cottevieille-Giraudet, R., 'Le *Catha edulis*: fut-il connu des Egyptiens?' *Institut Français d'Archéologie Orientale Bulletin* 35 (1935)

Crummey, Donald, 'Banditry and Resistance: Noble and Peasant in Nineteenth Century Ethiopia.' In Sven Rubenson, ed., *Proceedings of the Seventh International Conference of Ethiopian Studies.* Addis Ababa, East Lansing and Uppsala, 1984

Dercon, Stefan, and Lulseged Ayalew, 'Smuggling and Supply Response: Coffee in Ethiopia.' *World Development* 23 (1995)

Editorial, 'The Need for the Control of Khat.' *East African Medical Journal* 22 (1945)

Editorial, 'Poisoning by Khat or Mirra, *Catha edulis*,' *East African Medical Journal* 22 (1945)

Elmi, Abdulahi S., 'The Chewing of Khat in Somalia.' *Journal of Ethnopharmacology* 8 (1983)

Elmi, Abdulahi Sheikh, 'Khat: History, Spreading and Problems in Somalia.' In Thomas Labahn, ed., *Proceedings of the Second International Congress of Somali Studies.* Hamburg, 1984

Ferman, Louis A., Stuart Henry and Michèle Hoyman (special issue editors), *The Informal Economy: A Special Issue of the Annals of the American Academy of Political and Social Science* 493 (1987)

Fernyhough, Timothy, 'Social Mobility and Dissident Elites in Northern Ethiopia: The Role of Bandits, 1900–69.' In Donald Crummey, ed., *Banditry, Rebellion and Social Protest in Africa.* Portsmouth, 1986

Fresnel, F., 'L'Arabie vue en 1837–1838.' *Journal Asiatique* 5 (1871)

Friedmann, Harriet, 'Postscript: Small Commodity Production.' *Labor, Capital and Society* 19:1 (1986)

Gebre Wold Engeda Work, 'Ethiopia's Traditional System of Land Tenure and Taxation.' *Ethiopia Observer* 5 (1965) (Amharic)

Gershuny, J.I., and R.E. Pahl, 'The Future of the Informal Economy.' *New Society* 3 (1980)

Geshekter, Charles L., 'Anti-Colonialism and Class Formation: The Eastern Horn of Africa, 1920–1950.' In Thomas Labahn, ed., *Proceedings of the Second International Congress of Somali Studies.* Hamburg, 1984

Greenway, P.J., 'Khat', *East African Agricultural Journal* 13 (1947)

Grove, A.T., 'Population Densities and Agriculture in Northern Nigeria.' In K.M. Barbour and R.M. Prothero, eds, *Essays on African Population.* London, 1961

Halbach, H., 'Medical Aspects of the Chewing of Khat Leaves.' *Bulletin of the World Health Organization* 47 (1972)

Hart, Keith, 'Informal Income Opportunities and Urban Unemployment in Ghana.' *Journal of Modern African Studies* 11 (1973)

Heacock, R.A., and J.E. Forrest, 'Khat.' *Canadian Journal of Pharmaceutical Sciences* 3:9 (1974)

Heisch, R.B., 'A Case of Poisoning by *Catha edulis*.' *The East African Medical Journal* 22 (1945)

Henry, Stuart, 'The Working Unemployed: Perspectives on the Informal Economy and Unemployment.' *Sociological Review* 30 (1982)

Hess, J.J., 'Kat.' In *Encyclopedia of Islam.* London, 1976

Hill, Bob, 'Cat (*Catha edulis* Forsk.).' *Journal of Ethiopian Studies* 3 (1965)

Hjort, Anders, 'Trading Miraa: From School-leaver to Shop-owner in Kenya.' *Ethnos* 39 (1974)

Hobley, C.W., '*Catha edulis.*' *East African Natural History Society Journal* 2 (1912)

Hocking, G.M., '*Catha edulis* ('Kath' or 'Kat'): A Review.' *Quarterly Journal of Crude Drug Research* 3 (1963)

Isaacman, Allen F., 'Peasants and Rural Social Protest in Africa.' *African Studies Review* 33:2 (1990)

Johnson, A., 'Security and Risk Taking Among Poor Peasants: A Brazilian Case Study.' In George Dalton, ed., *Studies in Economic Anthropology*. Washington, D.C., 1971

Jones, Christine, and Michael Roemer, 'Editors' Introduction: Modeling and Measuring Parallel Markets in Developing Countries.' *World Development* 17:12 (1989)

Kalix, P., 'The Pharmacology of Khat and of the Khat Alkaloid Cathinone.' In B. Shahandeh, R. Geadah, A. Toungue, E. Toungue and J. Rolli, eds, *The Health and Socio-economic Effects of Khat Use: Proceedings of an International Conference on Khat.* Lausanne, 1983

Kalix, Peter, 'The Pharmacology of Khat.' In Ermias Dagne, ed., *Proceedings of the International Symposium on Khat (Catha edulis): Chemical and Ethnopharmacological Aspects.* Addis Ababa, 1984

Kennedy, J.G., J. Teague and L. Fairbanks, 'Qat Use in Yemen: A Study in Medical Anthropology.' *Culture, Medicine and Psychiatry* 4 (1980)

Kervingant, Dr, 'The Consumption of Khat in French Somaliland.' *Bulletin on Narcotics* 11 (1959)

Klaren, P.F., and T.J. Bossert, 'Lost Promise: Explaining Latin American Under-development.' In idem, *Promise of Development: Theories of Change in Latin America.* Boulder, 1986

Krikorian, Abraham, 'Kat and its Use: An Historical Perspective.' *Journal of Ethno-pharmacology* 12 (1984)

Lehmann, David, 'Two Paths of Agrarian Capitalism, or a Critique of Chaya-novian Marxism.' *Comparative Studies in Society and History* 28:4 (1986)

Lewis, I.M., 'Trade and Markets in Northern Somaliland.' In P. Bohannan and G. Dalton, eds, *Markets in Africa.* Evanston, Ill., 1965

Lindauer, David, 'Parallel, Fragmented, or Black? Defining Market Structure in Developing Economies.' *World Development* 17:12 (1989)

Lovejoy, Paul, 'Kola Nuts: The "Coffee" of the Central Sudan.' In Jordan Goodman, Paul E. Lovejoy and Andrew Sherratt, eds, *Consuming Habits: Drugs in History and Anthropology.* London, 1995

Luqman, W., and T.S. Danowski, 'The Use of Kat (*Catha edulis*) in Yemen. Social and Medical Observations.' *Annals of International Medicine* 85 (1976)

Mahteme Sellassie Wolde Maskal, 'The Land Tenure System of Ethiopia.' *Ethiopia Observer* 1 (1957)

Mancioli, M., and A. Parinello, 'Il qat (*Catha edulis*).' *La Clinica Terapeutica* 43, fasc 2: 269 (Ottobre 31, 1967)

Marcus, Harold, 'The Infrastructure of the Italo-Ethiopian Crisis: Haile Sellassie, the Solomonic Empire and the World Economy, 1916–1936.' In Robert Hess, ed., *Proceedings of the Fifth International Conference of Ethiopian Studies.* Chicago, 1978

Mazumdar, D. 'The Urban Informal Sector.' *World Development* 4 (1976)

McCann, James, 'The Political Economy of Rural Rebellion in Ethiopia: Northern Resistance to Imperial Expansion, 1928–1935.' *International Journal of African Historical Studies* 18:4 (1985)

McClellan, Charles, 'State Transformation and Social Reconstitution in Ethiopia: The Allure of the South.' *International Journal of African Historical Studies* 17:4 (1984)

McGee, T.G., 'Peasants in Cities: A Paradox, A Paradox, A Most Ingenious Paradox.' *Human Organization* 32 (1973)

McKee, C.M., 'Medical and Social Aspects of Qat in Yemen: A Review.' *Journal of the Royal Society of Medicine* 80 (1987)

MCTD, 'Our Coffee.' *A Publication of the Ministry of Coffee & Tea Development* (1988) (Amharic)

Merid Wolde Aregay, 'Millenarian Traditions and Peasant Movements in Ethiopia, 1500–1855.' In Sven Rubenson, ed., *Proceedings of the Seventh International Conference of Ethiopian Studies.* Addis Ababa, East Lansing and Uppsala, 1984

Mesfin Wolde-Mariam, 'An Estimate of the Population of Ethiopia.' *Ethiopia Observer* (1961)

Michaely, Michael, 'A Geometrical Analysis of Black Market Behavior.' *American Economic Review* 54 (1954)

Millar, James, 'A Reformulation of A.V. Chayanov's Theory of the Peasant Economy.' *Economic Development and Cultural Exchange* 8 (1970)

Ministry of Interior, 'Harerge Teqlay Gizat' (Harerge Province). *Annual Bulletin of the Ministry of Interior* (1953)

Mintz, Sidney, 'A Note on the Definition of Peasantries.' *Journal of Peasant Studies* 1 (1973)

Moktar, Mohammed, 'Notes sur le pays de Harrer.' *Bulletin de la Société Khédiviale de Géographie* 1:5 (1876)

Morris, Michael L., 'Official and Parallel Cereal Markets in Senegal: Empirical Evidence.' *World Development* 17:12 (1989)

Moser, Caroline O., 'Informal Sector or Petty Commodity Production: Dualism or Dependence in Urban Development.' *World Development* 6:9 (1978)

Moser, Charles, 'The Flower of Paradise: The Part which Khat Plays in the Life of the Yemen Arab.' *Geographical Journal* 93 (1917)

Owen, F., '*Catha edulis.*' *Journal of the Society of Chemical Industry* 29 (1910)

Pankhurst, E.S., 'The Beginnings of Modern Transport in Ethiopia.' *Ethiopia Observer* 1 (1958)

Pankhurst, E. Sylvia, 'Harer Under Egyptian Rule.' *Ethiopia Observer* 2:2 (1958)

Pankhurst, R.K.P., 'Ethiopia and the Red Sea and Gulf of Aden Ports in the Nineteenth and Twentieth Centuries.' *Ethiopia Observer* 8 (1964)

Pankhurst, R., 'Notes for the History of Ethiopian Agriculture.' *Ethiopia Observer* 7 (1964)

Pankhurst, R., 'A Chapter in Ethiopia's Commercial History: Developments during the Fascist Occupation.' *Ethiopia Observer* 14:1 (1971)

Paris, R., and H. Moyse, 'Abyssinian Tea (*Catha edulis* Forssk., Celastraceae).' *Bulletin on Narcotics* 10:2 (1958)

Paulitschke, P., 'Le Harrar sous l'Administration Egyptienne, 1875–1885.' *Bulletin de la Societé Khédiviale de Géographie d'Egypte* 10 (1887)

Peters, D., 'Khat: Its History, Botany, Chemistry and Toxicology.' *Pharmaceutical Journal* 169 (1952)

Pitts, Mark, 'Smuggling and Price Disparity.' *Journal of International Economics* 11 (1981)

Popenoe, P., '*Catha edulis*: A Narcotic of the Southern Arabs.' *Science* 43 (1916)

Poschen, P. 'An Evaluation of the *Acacia albida*-based Agroforestry Practices in the Harerghe Highlands of Eastern Ethiopia.' In P.K.R. Nair, ed., *Agroforestry Systems in the Tropics.* Dordrecht, 1989

Quenemoen, M.E., Darrell Watts and Yack Mosley, *The Economics of Irrigation with Small Pumps in the Chercher Highland.* Imperial Ethiopian College of Agriculture,

Experiment Station Bulletin 58 (Dire Dawa, 1968)

Radt, C., 'Contribution à l'histoire ethnographique d'une plante stimulante: Le Kat. Le Kat en Ethiopie.' *Ethnographie* 65 (1971)

Rahmato, Desalegn, 'Peasant Agriculture under the Old Regime.' In Shiferaw Bekele, ed., *Economic History of Modern Ethiopia, 1935–1974*. Vol. I. Oxford, 1995

Rodinson, Maxime, 'Esquisse d'une monographie du qat.' *Journal Asiatique* 265 (1977)

Roque, J. de la, *A Voyage to Arabia the Happy, By way of the Eastern Ocean, and the straits of the Red-Sea; Performed by the French for the first time, A.D. 1708, 1709, 1710*. London, 1726

Roseberry, William, 'The Ideology of Domestic Production.' *Labor, Capital and Society*, 19:1 (1983)

Rushby, Kevin, 'The High Life.' *Geographical Magazine* (January 1995)

Samatar, Abdi I., 'The Predatory State and the Peasantry: Reflections on Rural Development Policy in Somalia.' *Africa Today* 3rd Quarter (1985)

Sebhat Gebre Igziabher, 'Sheik Abadir's City.' *Yakatit* 16:1 (1992) (Amharic)

Sebsebe Demisew, 'Botanical Aspects of "Khat": *Catha Edulis* (Celastraceae).' In Ermias Dagne, ed., *Proceedings of the International Symposium on Khat (Catha edulis): Chemical and Ethnopharmacological Aspects*. Addis Ababa, 1984

Serjeant, R.B., 'The Market, Business Life, Occupations, the Legality and Sale of Stimulants.' In R.B. Serjeant and Ronald Lewcock, eds, *San'a: An Arabian Islamic City*. London, 1983

Shadan, Parirokh, and E.J. Shellard, 'An Anatomical Study of Ethiopian Khat (Leaf of *Catha edulis* Forsk.).' *Journal of Pharmacy and Pharmacology* 14 (1962)

Sheikh, Munir A. 'Black Market for Foreign Exchange, Capital Flows and Smuggling.' *Journal of Development Economics* 3 (1976)

Silverman, Sydel, 'The Peasant Concept in Anthropology.' *The Journal of Peasant Studies* 7:1 (1979)

Stockman, R., 'The Pharmacological Action of *Catha edulis* and its Alkaloids.' *Journal of Pharmacology and Experimental Therapeutics* 35 (1913)

Szendrei, Kalman, 'The Chemistry of Khat.' *Bulletin on Narcotics, Special Issue Devoted to Catha edulis (Khat)* 32:2 (1980)

Szendrei, Kalman, 'Recent Progress in Khat Chemistry.' In Ermias Dagne, ed., *Proceedings of the International Symposium on Khat (Catha edulis): Chemical and Ethnopharmacological Aspects*. Addis Ababa, 1984

Tesfaye Gebre Sellassie, 'The City of Harer and its Surroundings.' *Ethiopian Geographical Journal* 5:1 (1967) (Amharic)

Tesfaye Gebre Sellassie, 'Harer Awraja.' *Ethiopian Geographical Journal* 5:1 (1967)

Tipps, D.C., 'Modernization Theory and the Comparative Study of Societies: A Critical Perspective.' *Comparative Studies in Society and History* 15:2 (1973)

Trellu, Michel, 'The Pharmacodynamics of Khat.' *Bulletin on Narcotics* 11:2 (1959)

United Nations, 'The Problem of Khat.' Report of the Twelfth Session, April 29–May 31, 1957

UN Commission on Narcotic Drugs, 'Khat.' *Bulletin on Narcotics* 8 (1956)

UN Commission on Narcotic Drugs, '*Catha edulis* (Khat).' *Bulletin on Narcotics*. Special Issue devoted to *Catha edulis* (Khat) 32: 2 (1980)

UN Commission on Narcotic Drugs, 'Khat.' *Bulletin on Narcotics* 17:1 (1966)

UN Commission on Narcotic Drugs, 'The Question of Qat.' *Bulletin on Narcotics* 16:4 (1964)

UN Commission on Narcotic Drugs, 'The Medical and Social Problems of Khat in Djibouti.' *Bulletin on Narcotics* 9:4 (1957)

Varisco, Daniel Martin, 'On the Meaning of Chewing: The Significance of Qat in the Yemen Arab Republic.' *International Journal of Middle East Studies* 18 (1986)

Vaughan, James, 'Notes on the Drugs Observed at Aden, Arabia.' *Pharmaceutical Journal* 12:5 (1852)

Waldron, Sidney, 'The Political Economy of Harari–Oromo Relationships, 1559–1874.' *Northeast African Studies* 6:1–2 (1984)

Wharton, Clifton R., 'Risk, Uncertainty, and the Subsistence Farmer: Technological Innovation and Resistance to Change in the Context of Survival.' In George Dalton, ed., *Studies in Economic Anthropology*. Washington, D.C., 1971

WHO Advisory Group, 'Review of the Pharmacology of Khat.' *Bulletin on Narcotics, Special Issue Devoted to Catha edulis (Khat)* 12:3 (1980)

WHO Expert Report on Addiction-producing Drugs, 'Khat (*Catha edulis*).' *Bulletin on Narcotics* 16:2 (1964)

Yusuf Ahmed, 'An Inquiry into Some Aspects of the Economy of Harar and the Records of the Household Economy of the Amirs of Harar, 1825–1875.' *Ethnological Society Bulletin, University College of Addis Ababa* 10 (1960)

Newspapers

Al-Ayam, Editorial, Aden, October 26, 1960

Al-Qalam Al-Adani, Aden, A motion by a municipal councillor, February 13, 1963

Derfner, Larry, 'High on Gat: Gat Chewing, an Old Yemenite Custom, is Making a Mark on Israeli Society.' *The Jerusalem Post*, International Edition. November 16, 1996

Editorial, 'Despite War, Famine and Pestilence, the Khat Trade Thrives in East Africa.' *The New York Times*, December 14, 1995

Ethiopia Dimts (The Voice of Ethiopia), Amharic daily, October 5, 1961

'Khat has become the backbone of our country's economy.' *Moresh*, Amharic weekly, 93, January 24, 2000

Books

Allan, W., *The African Husbandman*. London, 1965

al-'Umari, Ibn Fadl Allah, *Masalik al-absar fi mamalik al-amsar*. Traduit et annoté avec une introduction par Maurice Gaudefoy-Demombynes. Paris, 1927

Aman, V.F., *Organization and Operation of a Marketing Cooperative in Ethiopia*. Imperial Ethiopian College of Agriculture and Mechanical Arts, Experiment Station Bulletin 43. Addis Ababa, 1965

Assefa Bequele and Eshetu Chole, *A Profile of the Ethiopian Economy*. Addis Ababa, 1969

Atsma Giorgis, *Atsma Giorgis and His Work 'History of the Galla and the Kingdom of Shoa.'* Translated and edited by Bairu Tafla. Stuttgart, 1987

Azias, R.P., and R. Chambard, *Cinque années de recherches archéologiques en Ethiopie*. Paris, 1931

Bartleet, E.J., *In the Land of Sheba*. Birmingham, 1934

Bates, Robert, *Rural Responses to Industrialization*. New Haven, 1976

Berry, Sara, *Fathers Work for their Sons: Accumulation, Mobility and Class Formation in an Extended Yoruba Community*. Berkeley, 1985

Berry, Sara, *No Condition is Permanent: The Social Dynamics of Agrarian Change in Sub-Saharan Africa*. Madison, 1993

Birke, Lakew, *A Credit Study of 100 Members of Alemaya Farmers Cooperative Society, 1965*. Imperial Ethiopian College of Agriculture and Mechanical Arts, Experiment Station Bulletin 46. Addis Ababa, 1966

Brotto, E., *Il regime della Terre nel Governo del Harar* (The Land Tenure System in the Government of Harer). Trans. Asmarom Legesse and Zerabrook Aberra (1966)

Bunker, Steven, *Peasants Against the State: The Politics of Market Control in Bugissu, Uganda, 1900–1983*. Champaign-Urbana, 1987

Burns, Scott, *The Household Economy*. Boston, 1977

Burton, Richard F., *First Footsteps in East Africa or, An Exploration of Harrar*. New York, 1987

Bury, Wyman, *Arabia Felix: Or the Turks in Yemen*. London, 1915

Clapham, Christopher, *Transformation and Continuity in Revolutionary Ethiopia*. Cambridge, 1988

Cohen, Abner, *Customs and Politics in Urban Africa: A Study of Hausa Migrants in Yoruba Towns*. Berkeley, 1969

Comyn-Platt, T., *The Abyssinian Storm*. London, 1935

Cook, S., and L. Binford, *Obliging Need: Rural Petty Industry in Mexican Capitalism*. Austin, 1990

Davis, K.C., Ahmed Mohammed and W.A. Wayt, *Farm Organization of Terre and Galmo Villages: Harer Province, Ethiopia*. Imperial Ethiopian College of Agriculture and Mechanical Arts, Experiment Station Bulletin 42. Addis Ababa, 1965

de Janvry, Alain, *The Agrarian Question and Reformism in Latin America*. Baltimore and London, 1981

Demissie Gebre Michael, *Land Tenure in Bate*. Imperial Ethiopian College of Agriculture, Experiment Station Bulletin 49. Dire Dawa, 1966

Douin, G., *Histoire du Règne Khédive Ismaïl*. Tome III. Cairo, 1941

Duchesne-Fournet, J., *Mission en Ethiopia, 1901–1903*. Paris, 1908–9

Farrago, L., *Arabian Antic*. New York, 1938

Gashaw, Mulugeta, and Minilik Tibebe Sellassie, *Ethiopian Village Studies, Hadele Keke, Kersa Woreda, Harerghe*. Harerghe, 1995

Gavin, R.J., *Aden Under British Rule, 1838–1967*. London, 1975

Geertz, Clifford, *Peddlers and Princes: Social Change and Economic Modernization in Two Indonesian Towns*. Chicago, 1963

Gilmour, T.L., *Abyssinia, the Ethiopian Railway and the Powers*. London, 1906

Gleichen, Count, *With the Mission to Menelik 1897*. London, 1971. First published in 1898

Goodman, D., and M. Redclift, *From Peasant to Proletariat: Capitalist Development and Agrarian Transitions*. New York, 1982

Griffin, Keith, ed., *The Economy of Ethiopia*. London, 1992

Grigg, David, *Population Growth and Agrarian Change: An Historical Perspective*. Cambridge, 1980

Gruhl, M., *The Citadel of Abyssinia*. London, 1932

Haberland, Eike, *Galla Sud-Æthiopiens*. Stuttgart, 1963

Harris, W. Cornwallis, *The Highlands of Aethiopia*. Vol. 3. London, 1844

Hirschman, Albert O., *Exit, Voice and Loyalty: Responses to Decline in Firms, Organizations and States*. Cambridge, Mass., 1970

Huffnagel, H.P., ed., *Agriculture in Ethiopia*. Rome, 1961

Huntingford, G.W.B., *The Galla of Ethiopia: The Kingdom of Kaffa and Janjero*. London, 1955

Huntingford, G.W.B., *The Glorious Victories of 'Amda Syon, King of Ethiopia*. Oxford, 1965

Ingrams, Harold, *Arabia and the Isles*. New York, 1966

Jennings, J.W., and C. Addison, *With the Abyssinians in Somaliland*. London, 1905

Jensen, P.C.M., *Spices, Condiments and Medicinal Plants in Ethiopia: Their Taxonomy and Agricultural Significance*. Wageningen, 1981

Kennedy, John G., *The Flower of Paradise: The Institutionalized Use of the Drug Qat in North Yemen*. Dordrecht, 1987

Lakew Birke, *A Credit Study of 100 Members of Alemaya Farmers Cooperative Society, 1965*. Imperial Ethiopian College of Agriculture and Mechanical Arts, Experiment Station Bulletin 46. Dire Dawa, 1966

Lefebvre, Théophile *et al.*, *Voyage en Abyssinie exécuté pendant les années 1839, 1841, 1842, 1843*. Paris, 1845–51

Legesse, Asmarom, *Gada: Three Approaches to the Study of an African Society*. New York, 1973

Le Roux, Hughes, *Ménélik et nous*. Paris, 1902

Leslau, Wolf, *Ethiopians Speak: Studies in Cultural Background, I. Harari*. Berkeley and Los Angeles, 1965

Lewis, I.M., *A Pastoral Democracy*. 2nd edn, London, 1982

Lovejoy, Paul, *Caravans of Kola: The Hausa Kola Trade, 1700–1900*. Zaria, 1980

McCann, James, *From Poverty to Famine in Northeast Ethiopia: A Rural History, 1900-1935*. Philadelphia, 1987

McCann, James, *People of the Plow: An Agricultural History of Ethiopia, 1800–1990*. Madison, 1995

Mengisteab, Kidane, *Ethiopia: Failure of Land Reform and Agricultural Crisis*. New York, 1990

Migdal, Joel S., *Strong States and Weak States: State–Society Relations and State Capabilities in the Third World*. Princeton, 1988

Miller, Leonard, and Telahun Makonnen, *Organization and Operation of Three Ethiopian Case Farms*. Imperial Ethiopian College of Agriculture, Experiment Station Bulletin 35. Dire Dawa, 1965

Nelhans, Bertil, *Khat: A Stimulating Drug in East Africa and the Arabian Peninsula*. Götenberg, 1974

Niebuhr, C., *Travels through Arabia and other Countries of the East*, translated into English by Robert Heron. Vol. 2. Dublin, 1792

Pankhurst, Richard, *Economic History of Ethiopia, 1800–1935*, Addis Ababa, 1968

Perham, M. *The Government of Ethiopia*. 2nd edn, London, 1969

Popkin, Samuel L. *The Rational Peasant: The Political Economy of Rural Society in Vietnam*. Berkeley and Los Angeles, 1979

Rahini, Amin, *Arabian Peak and Desert: Travels in al-Yaman*. Boston, 1930

Rahmato, Desalegn, *Agrarian Reform in Ethiopia*. Uppsala, 1984

Reveri, Raman, *Catha edulis Forssk.: Geographical Dispersal, Botanical, Ecological and Agronomical Aspects with Special Reference to Yemen Arab Republic*. Göttingen, 1983

Rey, C.F., *Unconquered Abyssinia*. London, 1923

Rihani, A., *Arabian Peak and Desert: Travels in al-Yemen*. Boston, 1930

Robecchi-Bricchetti, L., *Nell'Harrar*. Milano, 1903

Rushby, Kevin, *Eating the Flowers of Paradise: A Journey through the Drug Fields of Ethiopia and Yemen*. New York, 1999

Sabry, M., *Épisode de la Question d'Afrique: L'Empire Égyptien sous Ismaïl et l'Ingérence Anglo-Française, 1863–1879*. Paris, 1933

Sahlins, Marshal, *Stone Age Economics*. Chicago, 1972

Salt, Henry, *A Voyage to Abyssinia*. London, 1814

Samatar, Abdi Ismail, *The State and Rural Transformation in Northern Somalia, 1884–1986*. Madison, 1989

Sanabria, Harry, *The Coca Boom and Rural Social Change in Bolivia.* Ann Arbor, 1993

Santagata, F., *L'Harar: territorio di pace e di civilità.* Milano, 1940

Scarin, Emilio, *Hararino: Richerche e Studi Geografici.* Florence, 1942

Scott, James, *The Moral Economy of the Peasant: Rebellion and Subsistence in South East Asia.* New Haven, 1976

Seleshi Wolde-Tsadik, *Land Ownership in Harerge Province.* Imperial Ethiopian College of Agriculture, Experiment Station Bulletin 47. Dire Dawa, 1966

Seleshi Wolde-Tsadik, *Land Taxation in Harerge Province.* Imperial Ethiopian College of Agriculture, Experiment Station Bulletin 48. Dire Dawa, 1966

Skinner, R.P., *Abyssinia of Today.* London, 1906

Storck, Harmen *et al., Farming Systems and Farm Management Practices of Smallholders in the Hararghe Highlands.* Kiel, 1991

Thomas, J.J., *Informal Economic Activity.* Ann Arbor, 1992

Thompson, Virginia, and Richard Adloff, *Djibouti and the Horn of Africa.* Stanford, 1968

Trimingham, John Spencer, *Islam in Ethiopia.* London, 1952

Weir, Shelagh, *Qat in Yemen: Consumption and Social Change.* London, 1985

Wellby, M., *'Twixt Sirdar and Menelik.* London, 1901

Westphal, E., *Agricultural Systems in Ethiopia.* Wageningen, 1975

Williams, Neville, *Contraband Cargoes: Seven Centuries of Smuggling.* London, 1961

World Bank, *Ethiopia: Export Action Program.* Washington, D.C., 1987

Wyman, Bury, *Arabia Felix: Or the Turks in Yemen.* London, 1915

Zervos, Adrien, *L'Empire d'Ethiopie: le miroir de l'Ethiopie moderne, 1906–1935.* Alexandria, 1936

Index